Teacher Edition

Eureka Math Grade 7 Module 2

Special thanks go to the Gordon A. Cain Center and to the Department of Mathematics at Louisiana State University for their support in the development of *Eureka Math*.

For a free *Eureka Math* Teacher Resource Pack, Parent Tip Sheets, and more please visit www.Eureka.tools

Published by Great Minds

Printed in the U.S.A.
This book may be purchased from the publisher at eureka-math.org
10 9 8 7 6 5

ISBN 978-1-63255-614-1

Eureka Math: A Story of Ratios Contributors

Michael Allwood, Curriculum Writer
Tiah Alphonso, Program Manager—Curriculum Production
Catriona Anderson, Program Manager—Implementation Support
Beau Bailey, Curriculum Writer
Scott Baldridge, Lead Mathematician and Lead Curriculum Writer
Bonnie Bergstresser, Math Auditor
Gail Burrill, Curriculum Writer
Beth Chance, Statistician
Joanne Choi, Curriculum Writer
Jill Diniz, Program Director
Lori Fanning, Curriculum Writer
Ellen Fort, Math Auditor
Kathy Fritz, Curriculum Writer
Glenn Gebhard, Curriculum Writer
Krysta Gibbs, Curriculum Writer
Winnie Gilbert, Lead Writer / Editor, Grade 8
Pam Goodner, Math Auditor
Debby Grawn, Curriculum Writer
Bonnie Hart, Curriculum Writer
Stefanie Hassan, Lead Writer / Editor, Grade 8
Sherri Hernandez, Math Auditor
Bob Hollister, Math Auditor
Patrick Hopfensperger, Curriculum Writer
Sunil Koswatta, Mathematician, Grade 8
Brian Kotz, Curriculum Writer
Henry Kranendonk, Lead Writer / Editor, Statistics
Connie Laughlin, Math Auditor
Jennifer Loftin, Program Manager—Professional Development
Nell McAnelly, Project Director
Ben McCarty, Mathematician
Stacie McClintock, Document Production Manager
Saki Milton, Curriculum Writer
Pia Mohsen, Curriculum Writer
Jerry Moreno, Statistician
Ann Netter, Lead Writer / Editor, Grades 6–7
Sarah Oyler, Document Coordinator
Roxy Peck, Statistician, Lead Writer / Editor, Statistics
Terrie Poehl, Math Auditor
Kristen Riedel, Math Audit Team Lead
Spencer Roby, Math Auditor
Kathleen Scholand, Math Auditor
Erika Silva, Lead Writer / Editor, Grade 6–7
Robyn Sorenson, Math Auditor
Hester Sutton, Advisor / Reviewer Grades 6–7
Shannon Vinson, Lead Writer / Editor, Statistics
Allison Witcraft, Math Auditor

Julie Wortmann, Lead Writer / Editor, Grade 7
David Wright, Mathematician, Lead Writer / Editor, Grades 6–7

This page intentionally left blank

Table of Contents[1]

Rational Numbers

[1]Each lesson is ONE day, and ONE day is considered a 45-minute period.

Grade 7 • Module 2

Rational Numbers

OVERVIEW

In Grade 6, students formed a conceptual understanding of integers through the use of the number line, absolute value, and opposites and extended their understanding to include the ordering and comparing of rational numbers (**6.NS.C.5**, **6.NS.C.6**, **6.NS.C.7**). This module uses the Integer Game: a card game that creates a conceptual understanding of integer operations and serves as a powerful mental model students can rely on during the module. Students build on their understanding of rational numbers to add, subtract, multiply, and divide signed numbers. Previous work in computing the sums, differences, products, and quotients of fractions and decimals serves as a significant foundation as well.

In Topic A, students return to the number line to model the addition and subtraction of integers (**7.NS.A.1**). They use the number line and the Integer Game to demonstrate that an integer added to its opposite equals zero, representing the additive inverse (**7.NS.A.1a**, **7.NS.A.1b**). Their findings are formalized as students develop rules for adding and subtracting integers, and they recognize that subtracting a number is the same as adding its opposite (**7.NS.A.1c**). Real-life situations are represented by the sums and differences of signed numbers. Students extend integer rules to include the rational numbers and use properties of operations to perform rational number calculations without the use of a calculator (**7.NS.A.1d**).

Students develop the rules for multiplying and dividing signed numbers in Topic B. They use the properties of operations and their previous understanding of multiplication as repeated addition to represent the multiplication of a negative number as repeated subtraction (**7.NS.A.2a**). Students make analogies to the Integer Game to understand that the product of two negative numbers is a positive number. From earlier grades, they recognize division as the inverse process of multiplication. Thus, signed number rules for division are consistent with those for multiplication, provided a divisor is not zero (**7.NS.A.2b**). Students represent the division of two integers as a fraction, extending product and quotient rules to all rational numbers. They realize that any rational number in fractional form can be represented as a decimal that either terminates in 0s or repeats (**7.NS.A.2d**). Students recognize that the context of a situation often determines the most appropriate form of a rational number, and they use long division, place value, and equivalent fractions to fluently convert between these fractions and decimal forms. Topic B concludes with students multiplying and dividing rational numbers using the properties of operations (**7.NS.A.2c**).

In Topic C, students problem-solve with rational numbers and draw upon their work from Grade 6 with expressions and equations (**6.EE.A.2**, **6.EE.A.3**, **6.EE.A.4**, **6.EE.B.5**, **6.EE.B.6**, **6.EE.B.7**). They perform operations with rational numbers (**7.NS.A.3**), incorporating them into algebraic expressions and equations. They represent and evaluate expressions in multiple forms, demonstrating how quantities are related (**7.EE.A.2**). The Integer Game is revisited as students discover *if-then* statements, relating changes in player's hands (who have the same card-value totals) to changes in both sides of a number sentence. Students translate word problems into algebraic equations and become proficient at solving equations of the form $px + q = r$ and $p(x + q) = r$, where p, q, and r, are specific rational numbers (**7.EE.B.4a**). As they become fluent in generating algebraic solutions, students identify the operations, inverse operations, and order of steps, comparing these to an arithmetic solution. Use of algebra to represent contextual problems continues in Module 3.

This module is comprised of 23 lessons; 7 days are reserved for administering the Mid- and End-of-Module Assessments, returning the assessments, and remediating or providing further applications of the concepts. The Mid-Module Assessment follows Topic B, and the End-of-Module Assessment follows Topic C.

Focus Standards

Apply and extend previous understandings of operations with fractions to add, subtract, multiply, and divide rational numbers.

7.NS.A.1 Apply and extend previous understandings of addition and subtraction to add and subtract rational numbers; represent addition and subtraction on a horizontal or vertical number line diagram.

a. Describe situations in which opposite quantities combine to make 0. *For example, a hydrogen atom has 0 charge because its two constituents are oppositely charged.*

b. Understand $p + q$ as the number located a distance $|q|$ from p, in the positive or negative direction depending on whether q is positive or negative. Show that a number and its opposite have a sum of 0 (are additive inverses). Interpret sums of rational numbers by describing real-world contexts.

c. Understand subtraction of rational numbers as adding the additive inverse, $p - q = p + (-q)$. Show that the distance between two rational numbers on the number line is the absolute value of their difference, and apply this principle in real-world contexts.

d. Apply properties of operations as strategies to add and subtract rational numbers.

7.NS.A.2 Apply and extend previous understandings of multiplication and division and of fractions to multiply and divide rational numbers.

a. Understand that multiplication is extended from fractions to rational numbers by requiring that operations continue to satisfy the properties of operations, particularly the distributive property, leading to products such as $(-1)(-1) = 1$ and the rules for multiplying signed numbers. Interpret products of rational numbers by describing real-world contexts.

b. Understand that integers can be divided, provided that the divisor is not zero, and every quotient of integers (with non-zero divisor) is a rational number. If p and q are integers, then $-(p/q) = (-p)/q = p/(-q)$. Interpret quotients of rational numbers by describing real-world contexts.

c. Apply properties of operations as strategies to multiply and divide rational numbers.

d. Convert a rational number to a decimal using long division; know that the decimal form of a rational number terminates in 0s or eventually repeats.

7.NS.A.3 Solve real-world and mathematical problems involving the four operations with rational numbers.[2]

Use properties of operations to generate equivalent expressions.

7.EE.A.2[3] Understand that rewriting an expression in different forms in a problem context can shed light on the problem and how the quantities in it are related. *For example,* $a + 0.05a = 1.05a$ *means that "increase by* 5%*" is the same as "multiply by* 1.05*."*

Solve real-life and mathematical problems using numerical and algebraic expressions and equations.

7.EE.B.4[4] Use variables to represent quantities in a real-world or mathematical problem, and construct simple equations and inequalities to solve problems by reasoning about the quantities.

a. Solve word problems leading to equations of the form $px + q = r$ and $p(x + q) = r$, where p, q, and r are specific rational numbers. Solve equations of these forms fluently. Compare an algebraic solution to an arithmetic solution, identifying the sequence of the operations used in each approach. *For example, the perimeter of a rectangle is* 54 cm. *Its length is* 6 cm. *What is its width?*

Foundational Standards

Use equivalent fractions as a strategy to add and subtract fractions.

5.NF.A.1 Add and subtract fractions with unlike denominators (including mixed numbers) by replacing given fractions with equivalent fractions in such a way as to produce an equivalent sum or difference of fractions with like denominators. *For example,* $2/3 + 5/4 = 8/12 + 15/12 = 23/12$. *(In general,* $a/b + c/d = (ad + bc)/bd$*.)*

Apply and extend previous understandings of multiplication and division to multiply and divide fractions.

5.NF.B.3 Interpret a fraction as division of the numerator by the denominator ($a/b = a \div b$). Solve word problems involving division of whole numbers leading to answers in the form of fractions or mixed numbers, e.g., by using visual fraction models or equations to represent the problem. *For example, interpret* $3/4$ *as the result of dividing* 3 *by* 4*, noting that* $3/4$ *multiplied by* 4 *equals* 3*, and that when* 3 *wholes are shared equally among* 4 *people each person has a share of size* $3/4$*. If* 9 *people want to share a* 50*-pound sack of rice equally by weight, how many pounds of rice should each person get? Between what two whole numbers does your answer lie?*

---

[2] Computations with rational numbers extend the rules for manipulating fractions to complex fractions.
[3] In this module, this standard is applied to expressions with rational numbers in them.
[4] In this module, the equations include negative rational numbers.

5.NF.B.4 Apply and extend previous understandings of multiplication to multiply a fraction or whole number by a fraction.

a. Interpret the product $(a/b) \times q$ as a parts of a partition of q into b equal parts; equivalently, as the result of a sequence of operations $a \times q \div b$. *For example, use a visual fraction model to show* $(2/3) \times 4 = 8/3$, *and create a story context for this equation. Do the same with* $(2/3) \times (4/5) = 8/15$. *(In general,* $(a/b) \times (c/d) = ac/bd$.*)*

Apply and extend previous understandings of multiplication and division to divide fractions by fractions.

6.NS.A.1 Interpret and compute quotients of fractions, and solve word problems involving division of fractions by fractions, e.g., by using visual fraction models and equations to represent the problem. *For example, create a story context for* $(2/3) \div (3/4)$ *and use a visual fraction model to show the quotient; use the relationship between multiplication and division to explain that* $(2/3) \div (3/4) = 8/9$ *because* $3/4$ *of* $8/9$ *is* $2/3$. *(In general,* $(a/b) \div (c/d) = ad/bc$.*) How much chocolate will each person get if* 3 *people share* $1/2$ lb *of chocolate equally? How many* $3/4$*-cup servings are in* $2/3$ *of a cup of yogurt? How wide is a rectangular strip of land with length* $3/4$ mi *and area* $1/2$ *square mi?*

Compute fluently with multi-digit numbers and find common factors and multiples.

6.NS.B.3 Fluently add, subtract, multiply, and divide multi-digit decimals using the standard algorithm for each operation.

Apply and extend previous understandings of numbers to the system of rational numbers.

6.NS.C.5 Understand that positive and negative numbers are used together to describe quantities having opposite directions or values (e.g., temperature above/below zero, elevation above/below sea level, credits/debits, positive/negative electric charge); use positive and negative numbers to represent quantities in real-world contexts, explaining the meaning of 0 in each situation.

6.NS.C.6 Understand a rational number as a point on the number line. Extend number line diagrams and coordinate axes familiar from previous grades to represent points on the line and in the plane with negative number coordinates.

a. Recognize opposite signs of numbers as indicating locations on opposite sides of 0 on the number line; recognize that the opposite of the opposite of a number is the number itself, e.g., $-(-3) = 3$, and that 0 is its own opposite.

6.NS.C.7 Understand ordering and absolute value of rational numbers.

c. Understand the absolute value of a rational number as its distance from 0 on the number line; interpret absolute value as magnitude for a positive or negative quantity in a real-world situation. *For example, for an account balance of* -30 *dollars, write* $|-30| = 30$ *to describe the size of the debt in dollars.*

Apply and extend previous understandings of arithmetic to algebraic expressions.

6.EE.A.2 Write, read, and evaluate expressions in which letters stand for numbers.

a. Write expressions that record operations with numbers and with letters standing for numbers. *For example, express the calculation "Subtract y from* 5*" as* $5 - y$.

b. Identify parts of an expression using mathematical terms (sum, term, product, factor quotient, coefficient); view one or more parts of an expression as a single entity. *For example, describe the expression* $2\ (8 + 7)$ *as a product of two factors; view* $(8 + 7)$ *as both a single entity and a sum of two terms.*

c. Evaluate expressions at specific values of their variables. Include expressions that arise from formulas used in real-world problems. Perform arithmetic operations, including those involving whole-number exponents, in the conventional order when there are no parentheses to specify a particular order (Order of Operations). *For example, use the formulas* $V = s^3$ *and* $A = 6\ s^2$ *to find the volume and surface area of a cube with sides of length* $s = 1/2$.

6.EE.A.3 Apply the properties of operations to generate equivalent expressions. *For example, apply the distributive property to the expression* $3\ (2 + x)$ *to produce the equivalent expression* $6 + 3x$*; apply the distributive property to the expression* $24x + 18y$ *to produce the equivalent expression* $6\ (4x + 3y)$*; apply properties of operations to* $y + y + y$ *to produce the equivalent expression* $3y$.

6.EE.A.4 Identify when two expressions are equivalent (i.e., when the two expressions name the same number regardless of which value is substituted into them). *For example, the expressions* $y + y + y$ *and* $3y$ *are equivalent because they name the same number regardless of which number* y *stands for.*

Reason about and solve one-variable equations and inequalities.

6.EE.B.6 Use variables to represent numbers and write expressions when solving a real-world or mathematical problem; understand that a variable can represent an unknown number, or, depending on the purpose at hand, any number in a specified set.

6.EE.B.7 Solve real-world and mathematical problems by writing and solving equations of the form $x + p = q$ and $px = q$ for cases in which p, q and x are all nonnegative rational numbers.

Focus Standards for Mathematical Practice

MP.1 **Make sense of problems and persevere in solving them**. When problem-solving, students use a variety of techniques to make sense of a situation involving rational numbers. For example, they may draw a number line and use arrows to model and make sense of an integer addition or subtraction problem. Or when converting between forms of rational numbers, students persevere in carrying out the long division algorithm to determine a decimal's repeat pattern. A tape diagram may be constructed as an entry point to make sense of a working-backward problem. As students fluently solve word problems using algebraic equations and inverse operations, they consider their steps and determine whether or not they make sense in relationship to the arithmetic reasoning that served as their foundation in earlier grades.

MP.2 **Reason abstractly and quantitatively**. Students make sense of integer addition and subtraction through the use of an integer card game and diagramming the distances and directions on the number line. They use different properties of operations to add, subtract, multiply, and divide rational numbers, applying the properties to generate equivalent expressions or explain a rule. Students use integer subtraction and absolute value to justify the distance between two numbers on the number line. Algebraic expressions and equations are created to represent relationships. Students know how to use the properties of operations to solve equations. They make *zeros and ones* when solving an algebraic equation, thereby demonstrating an understanding of how the use of inverse operations ultimately leads to the value of the variable.

MP.4 **Model with mathematics**. Through the use of number lines, tape diagrams, expressions, and equations, students model relationships between rational numbers. Students relate operations involving integers to contextual examples. For instance, an overdraft fee of $\$25$ that is applied to an account balance of $-\$73.06$ is represented by the expression $-73.06 - 25$ or $-73.06 + (-25)$ using the additive inverse. Students compare their answers and thought processes in the Integer Game and use number line diagrams to ensure accurate reasoning. They deconstruct a difficult word problem by writing an equation, drawing a number line, or drawing a tape diagram to represent quantities. To find a change in elevation, students may draw a picture representing the objects and label their heights to aid in their understanding of the mathematical operation(s) that must be performed.

MP.6 **Attend to precision**. In performing operations with rational numbers, students understand that the decimal representation reflects the specific place value of each digit. When converting fractions to decimals, they carry out their calculations to specific place values, indicating a terminating or repeating pattern. In stating answers to problems involving signed numbers, students use integer rules and properties of operations to verify that the sign of their answer is correct. For instance, when finding an average temperature for temperatures whose sum is a negative number, students realize that the quotient must be a negative number since the divisor is positive and the dividend is negative.

MP.7 **Look for and make use of structure.** Students formulate rules for operations with signed numbers by observing patterns. For instance, they notice that adding -7 to a number is the same as subtracting 7 from the number, and thus, they develop a rule for subtraction that relates to adding the inverse of the subtrahend. Students use the concept of absolute value and subtraction to represent the distance between two rational numbers on a number line. They use patterns related to the properties of operations to justify the rules for multiplying and dividing signed numbers. The order of operations provides the structure by which students evaluate and generate equivalent expressions.

Terminology

New or Recently Introduced Terms

- **Additive Identity** (The additive identity is the number 0.)
- **Additive Inverse** (An *additive inverse of a number* is a number such that the sum of the two numbers is 0.
  The additive inverse of a number a is the opposite of a (i.e., $-a$) because if we add both numbers using the vector approach above, we get $a + (-a) = 0$.)
- **Formula for the Distance Between Two Numbers** (If p and q are numbers on a number line, then the *distance between* p *and* q is $|p - q|$.)
- **Multiplicative Identity** (The *multiplicative identity* is the number 1.)
- **Repeating Decimal Expansion** (A decimal expansion is *repeating* if, after some digit to the right of the decimal point, there is a finite string of consecutive digits called a block after which the decimal expansion consists entirely of consecutive copies of that block repeated forever.

  A common notation for a repeating decimal expansion is to write out the decimal expansion until the end of the first block and then writing a line over the block. For example, $3.1\overline{25}$ is a compact way to write the repeating decimal expansion $3.125252525....$)
- **Terminating Decimal Expansion** (A *terminating decimal expansion* is a repeating decimal expansion with period 1 and repeating digit 0. For example, the terminating decimal expansion of $\frac{1}{4}$ is $0.25000...$, or $0.25\overline{0}$.

  A number with a terminating decimal expansion can be written as a finite decimal. For example, the terminating decimal expansion $0.25000...$ represents the same number as the finite decimal 0.25.)

Familiar Terms and Symbols[5]

- Absolute Value
- Associative Property (of Multiplication and Addition)
- Commutative Property (of Multiplication and Addition)
- Credit
- Debit
- Deposit
- Distributive Property (of Multiplication Over Addition)
- Equation
- Expression
- Integer
- Inverse
- Multiplicative Inverse
- Negatives
- Opposites
- Overdraft
- Positives
- Rational Numbers
- Withdraw

Suggested Tools and Representations

- Equations
- Expressions
- Integer Game (See explanation on page 11)
- Number Line
- Tape Diagram

[5]These are terms and symbols students have seen previously.

Sprints

Sprints are designed to develop fluency. They should be fun, adrenaline-rich activities that intentionally build energy and excitement. A fast pace is essential. During Sprint administration, teachers assume the role of athletic coaches. A rousing routine fuels students' motivation to do their personal best. Student recognition of increasing success is critical, and so every improvement is acknowledged. (See the Sprint Delivery Script for the suggested means of acknowledging and celebrating student success.)

One Sprint has two parts with closely-related problems on each. Students complete the two parts of the Sprint in quick succession with the goal of improving on the second part, even if only by one more.

Sprints are not to be used for a grade. Thus, there is no need for students to write their names on the Sprints. The low-stakes nature of the exercise means that even students with allowances for extended time can participate. If a student finds the experience undesirable, it is recommended that the student be allowed to opt-out and take the Sprint home. In this case, regularly encourage the student to opt back in to participate in class.

With practice, the Sprint routine takes about 8 minutes.

Sprint Delivery Script

Gather the following: stopwatch, a copy of Sprint A for each student, a copy of Sprint B for each student, answers for Sprint A and Sprint B. The following delineates a script for delivery of a pair of Sprints.

This sprint covers: *topic.*

Do not look at the Sprint; keep it turned face down on your desk.

There are xx problems on the Sprint. You will have 60 seconds. Do as many as you can. I do not expect any of you to finish.

On your mark, get set, GO.

60 seconds of silence.

STOP. Circle the last problem you completed.

I will read the answers. You say "YES" if your answer matches. Mark the ones you have wrong. Don't try to correct them.

Energetically, rapid-fire call the answers ONLY.

Stop reading answers after there are no more students answering, "Yes."

Fantastic! Count the number you have correct, and write it on the top of the page. This is your personal goal for Sprint B.

Raise your hand if you have 1 or more correct. 2 or more, 3 or more, ...

Let us all applaud our runner up, [insert name], with x correct. And let us applaud our winner, [insert name], with x correct.

You have a few minutes to finish up the page and get ready for the next Sprint.

Students are allowed to talk and ask for help; let this part last as long as most are working seriously.

Stop working. I will read the answers again so you can check your work. You say "YES" if your answer matches.

Energetically, rapid-fire call the answers ONLY.

Optionally, ask students to stand and lead them in an energy-expanding exercise that also keeps the brain going. Examples are jumping jacks or arm circles, etc. while counting by 15*'s starting at* 15*, going up to* 150 *and back down to* 0*. You can follow this first exercise with a cool down exercise of a similar nature, such as calf raises with counting by one-sixths* $\left(\frac{1}{6}, \frac{1}{3}, \frac{1}{2}, \frac{2}{3}, \frac{5}{6}, 1, \dots\right)$.

Hand out the second Sprint and continue reading the script.

Keep the Sprint face down on your desk.

There are xx problems on the Sprint. You will have 60 seconds. Do as many as you can. I do not expect any of you to finish.

On your mark, get set, GO.

60 seconds of silence.

STOP. Circle the last problem you completed.

I will read the answers. You say "YES" if your answer matches. Mark the ones you have wrong. Don't try to correct them.

Quickly read the answers ONLY.

Count the number you have correct, and write it on the top of the page.

Raise your hand if you have 1 or more correct. 2 or more, 3 or more, ...

Let us all applaud our runner up, [insert name], with x correct. And let us applaud our winner, {insert name], with x correct.

Write the amount by which your score improved at the top of the page.

Raise your hand if you improved your score by 1 or more. 2 or more, 3 or more, ...

Let us all applaud our runner up for most improved, [insert name]. And let us applaud our winner for most improved, [insert name].

You can take the Sprint home and finish it if you want.

Assessment Summary

Assessment Type	Administered	Format	Standards Addressed
Mid-Module Assessment Task	After Topic B	Constructed response with rubric	7.NS.A.1, 7.NS.A.2
End-of-Module Assessment Task	After Topic C	Constructed response with rubric	7.NS.A.1, 7.NS.A.2, 7.NS.A.3, 7.EE.A.2, 7.EE.B.4a

Integer Game

Description

The Integer Game is a card game used throughout Grade 7 Module 2 to help students develop a conceptual understanding of integer operations. Game-play and rules can be adapted to meet the needs of a specific lesson objective or topic standard. This description of how to play the game sets the basis for use in the lessons.

How to Play

The Integer Game is designed for 2 to 4 players. Students play the game with a learning partner or with a cooperative learning group of students. Each player begins the game with a score of zero. The object of the game is to return to a score of zero by picking up and discarding integer cards. The number of cards dealt to each player can be adjusted based on students' familiarity with an operation and to differentiate for varying student ability levels. Below are the basic rules:

1. A student serves as the dealer (as well as a player) and provides each player with four cards.
2. The dealer turns one more card face up on the playing surface, starting a discard pile. The remaining cards become a draw pile.
3. The player to the dealer's left begins play. On his turn, a player may select the top card from either the draw pile or the discard pile. The player must keep this card and discard another card from his hand to the discard pile.
4. A player's goal is to have his hand's total card value stay as close to zero as possible. So for each turn, a player must determine how the card drawn affects his hand's total card value, by counting up or down accordingly. Also, a player must decide which card to discard so as to keep the total value of his hand as close to zero as possible.
5. Play continues with the next player, in the same manner, until all players have picked up and discarded a card four times.
6. The player(s) with a score of zero (or the closest to zero, as in Lesson 2) wins the round.

How the Integer Game Is Used in the Lessons

Lesson 1: Students try to reach a score of zero by obtaining the same number of positive points as negative points. This can be done by obtaining cards that are *additive inverses* or by obtaining combinations of cards that total opposite values. Students' prior work with recognizing and identifying numbers' opposites in Grade 6 serves as the basis for the extension to the addition of integers in this lesson.

Lesson 2: Students in this lesson start totaling their cards' values by using the number line and vectors as modeling tools to combine the values of positive and negative numbers. Players may win a round in this lesson if they obtain a score of zero or if they are the player whose score is closest to zero. The game-play and number line modeling fosters a conceptual understanding of absolute value as both distance (on the number line) and magnitude with regard to the amount by which a player's total point value is over or under zero.

Lesson 3: The Integer Game is used as a point of reference in Lesson 3 as the addition of integers becomes formalized.

Lesson 4: The Integer Game is again used as a point of reference in Lesson 4. Its simulation is used by students to justify the rules for adding integers.

Lesson 5: Students examine how picking up (adding) integer cards and laying down (subtracting) integer cards affects their score. They know that from earlier game-play that adding a positive value increases their score while adding a negative value decreases their score. Students also recognize that laying a card down is the opposite of picking a card up, so laying a card down represents subtraction. They understand that when a positive value is taken out of their hand their score decreases, but when a negative value is taken out of their hand their score increases. This serves as the basis for students' conceptual understanding of subtraction as adding the opposite.

Lesson 10: Students consider scenarios involving multiple sets of cards. They understand that picking up multiple cards of the same value is repeated addition of that value, and when the value is negative, it is the same as repeated subtraction of that value. They realize that laying down multiple negative cards (the opposite move) represents multiplying a negative integer by a negative integer. Examining these scenarios supports the development of the rules for multiplying integers (and eventually all signed numbers) in Lesson 11.

Lesson 11: The Integer Game is used as a point of reference in Lesson 11 as students use various scenarios as described in Lesson 10 to justify the rules for multiplying signed numbers.

Lesson 12: The Integer Game is revisited to model properties of equality (using if-then statements). Students use sets of cards with the same total score but different card values to explore what happens to the scores when equal values are added, subtracted, multiplied, or even divided from each of the hands.

The Integer Cards

1	2	3	4
5	6	7	8
9	10	11	12

−1	−2	−3	−4
−5	−6	−7	−8
−9	−10	−11	−12

0	0

Scaffolding Ideas for Diverse Learners

- Include a number line representation on the cards.

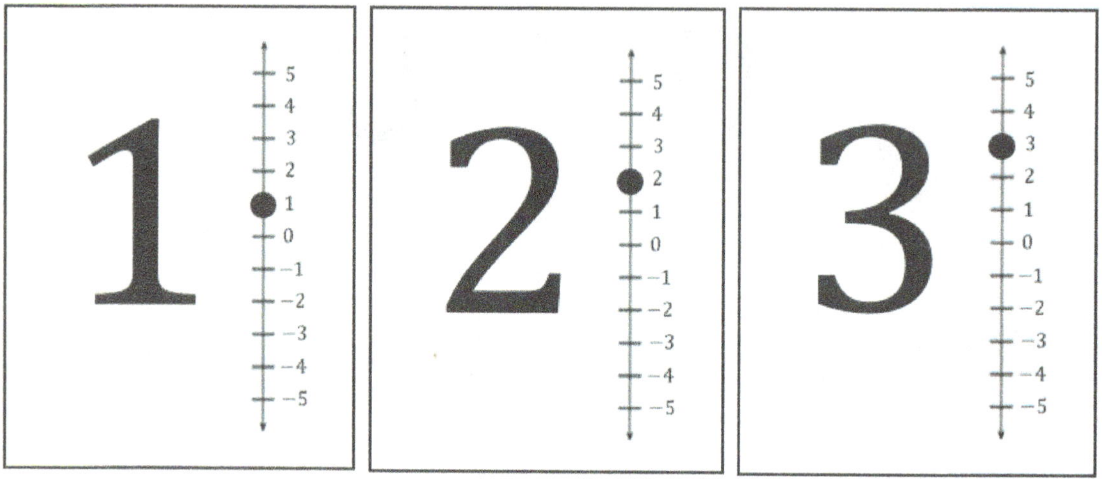

- Include counters on the cards.

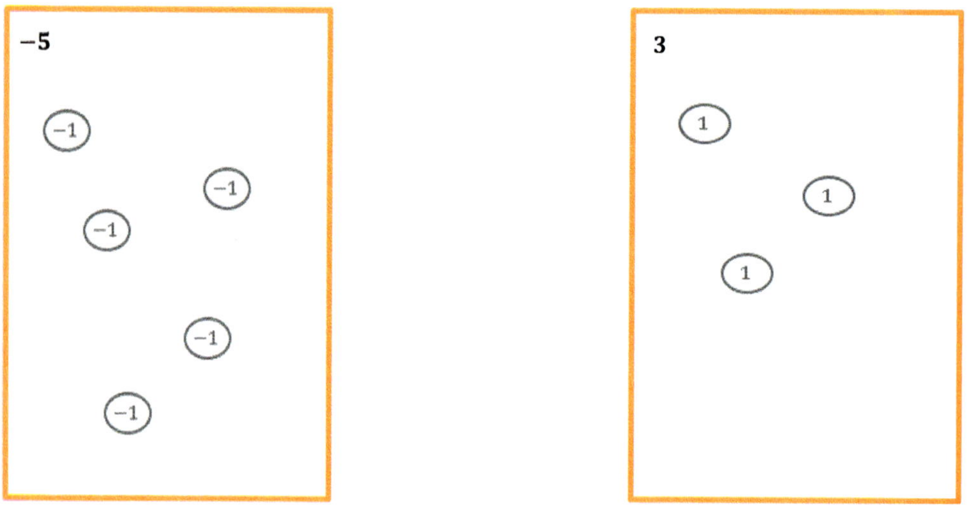

EUREKA MATH

Topic A

Addition and Subtraction of Integers and Rational Numbers

7.NS.A.1

Focus Standard: 7.NS.A.1 Apply and extend previous understandings of addition and subtraction to add and subtract rational numbers; represent addition and subtraction on a horizontal or vertical number line diagram.

a. Describe situations in which opposite quantities combine to make 0. *For example, a hydrogen atom has* 0 *charge because its two constituents are oppositely charged.*

b. Understand $p + q$ as the number located a distance $|q|$ from p, in the positive or negative direction depending on whether q is positive or negative. Show that a number and its opposite have a sum of 0 (are additive inverses). Interpret sums of rational numbers by describing real-world contexts.

c. Understand subtraction of rational numbers as adding the additive inverse, $p - q = p + (-q)$. Show that the distance between two rational numbers on the number line is the absolute value of their difference, and apply this principle in real-world contexts.

d. Apply properties of operations as strategies to add and subtract rational numbers.

Instructional Days: 9

Lesson 1: Opposite Quantities Combine to Make Zero (P)[1]

Lesson 2: Using the Number Line to Model the Addition of Integers (P)

Lesson 3: Understanding Addition of Integers (P)

Lesson 4: Efficiently Adding Integers and Other Rational Numbers (P)

Lesson 5: Understanding Subtraction of Integers and Other Rational Numbers (S)

---

[1]Lesson Structure Key: **P**-Problem Set Lesson, **M**-Modeling Cycle Lesson, **E**-Exploration Lesson, **S**-Socratic Lesson

In Topic A, students find sums and differences of signed numbers and establish rules related to the addition and subtraction of rational numbers (**7.NS.A.1**). Students draw upon experiences in modeling, ordering, and comparing integers and other rational numbers from Grade 6 Module 3 (**6.NS.C.5**, **6.NS.C.6**, **6.NS.C.7**). They use their previous work with adding and subtracting fractions and decimals (**5.NF.A.1**, **6.NS.B.3**) to compute the sums and differences of rational numbers. In Lesson 1, students play a card game called the Integer Game to understand how a number and its opposite combine to make zero. The number line is used to count up and down, serving as a visual model for finding sums. In Lessons 2 and 3, students more formally develop their understanding of the addition of integers. They use vectors to represent integers on the number line and apply the concept of absolute value (**6.NS.C.7c**) to represent the length of the vector while interpreting the sign of the integer as the vector's direction. By Lesson 4, students are efficiently adding integers using well-defined rules.

After addition rules are formalized, students begin subtracting integers in Lesson 5. They relate subtraction to removing a card from their hand in the Integer Game, realizing that subtracting a positive card has the same effect as adding or picking up a negative card. Similarly, removing (subtracting) a negative card increases students' scores the same way as adding the corresponding positive card. Therefore, students determine that subtracting a signed number is the same as adding its opposite. In Lesson 6, students deepen their understanding of subtraction using absolute value and the number line to justify that the distance between two signed numbers is the absolute value of their difference. They represent sums and differences of rational numbers using the number line in Lesson 7 and use vectors to model the sum, $p + q$, or the difference, $p - q$. As Topic A concludes, students apply the properties of operations to add and subtract rational numbers in Lessons 8 and 9. Using the properties of operations and their fluency in adding and subtracting decimals and fractions from earlier grades, they rewrite numerical expressions in different forms to efficiently find sums and differences of signed numbers without the use of a calculator.

Mid-Module Assessment problems 1, 2, 4, and 6 may be administered at the conclusion of Topic A to serve as an intermediate assessment before students are introduced to Topic B.

Lesson 1: Opposite Quantities Combine to Make Zero

Student Outcomes

- Students add positive integers by counting up and negative integers by counting down (using curved arrows on the number line).
- Students play the Integer Game to combine integers, justifying that an integer plus its opposite add to zero.
- Students know the opposite of a number is called the additive inverse because the sum of the two numbers is zero.

Lesson Notes

There is some required preparation from teachers before the lesson begins. It is suggested that number lines are provided for students. However, it is best if students can reuse these number lines by having them laminated and using dry erase markers. Also, the Integer Game is used during this lesson, so the teacher should prepare the necessary cards for the game.

Classwork

Exercise 1 (3 minutes): Positive and Negative Numbers Review

In pairs, students will discuss "What I Know" about positive and negative integers to access prior knowledge. Have them record and organize their ideas in the graphic organizer in the student materials. At the end of discussion, the teacher will choose a few pairs to share out with the class.

Exercise 1: Positive and Negative Numbers Review

With your partner, use the graphic organizer below to record what you know about positive and negative numbers. Add or remove statements during the whole-class discussion.

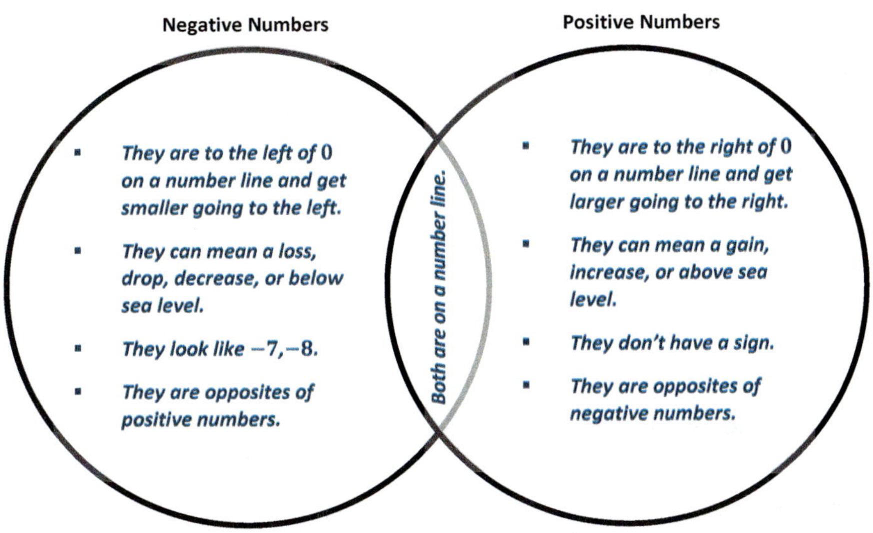

Scaffolding:

- Laminate (or place in sheet protectors) 1-page of number lines (vary blank and numbered) for individual use with dry erase markers.
- Create a number line on the floor using painters tape to model the counting on principle.
- Provide a wall model of the number line at the front of the room for visual reinforcement.

Example 1 (5 minutes): Introduction to the Integer Game

Read the Integer Game outline before the lesson. The teacher selects a group of 3 or 4 students to demonstrate to the whole class how to play the Integer Game.[1] The game will be played later in the lesson. The teacher should stress that the object of the game is to get a score of zero.

Example 2 (5 minutes): Counting Up and Counting Down on the Number Line

Model a few examples of counting with small curved arrows to locate numbers on the number line, where *counting up* corresponds to positive numbers and *counting down* corresponds to negative numbers.

Example 2: Counting Up and Counting Down on the Number Line

Use the number line below to practice counting up and counting down.

- ***Counting up starting at* 0 corresponds to ___*positive*___ numbers.**
- ***Counting down starting at* 0 corresponds to ___*negative*___ numbers.**

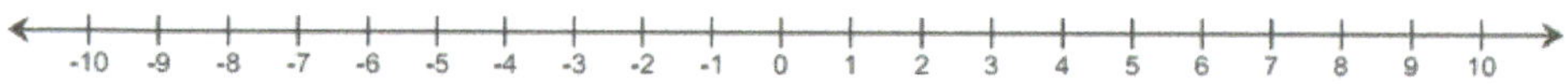

A negative 7 is 7 units to the left of 0. $|-7| = 7$

A positive 7 is 7 units to the right of 0. $|7| = 7$

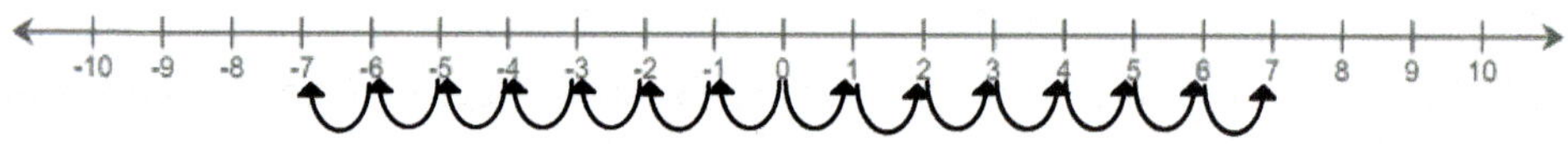

a. **Where do you begin when locating a number on the number line?**

Start at 0.

b. **What do you call the distance between a number and 0 on a number line?**

The absolute value

c. **What is the relationship between 7 and -7?**

Answers will vary. 7 and -7 both have the same absolute values. They are both the same distance from zero, 0, but in opposite directions; therefore, 7 and -7 are opposites.

---

[1]Refer to the Integer Game outline for player rules.

Example 3 (5 minutes): Using the Integer Game and the Number Line

The teacher leads the whole class using a number line to model the concept of counting on (addition) to calculate the value of a hand when playing the Integer Game. The hand's value is the sum of the card values.

5	-5	-4	8
First card: 5	Second Card: -5	Third Card: -4	Fourth Card: 8
Start at 0 and end up at positive 5. This is the first card drawn, so the value of the hand is 5.	Start at 5, the value of the hand after the first card; move 5 units to the left to end at 0.	Start at 0, the value of the hand after the second card; move 4 units to the left.	Start at -4, the value of the hand after the third card; move 8 units to the right.

- What is the final position on the number line?
  - *The final position on the number line is* 4.

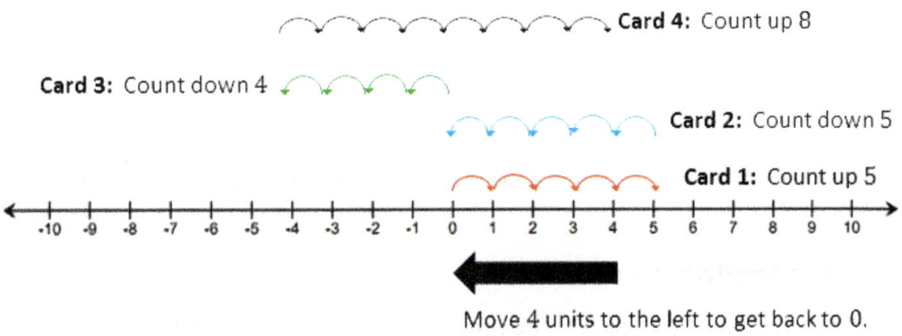

- What card or combination of cards would you need to get back to 0?
  - *In order to get a score of* 0*, I would need to count down* 4 *units. This means, I would need to draw a* -4 *card or a combination of cards whose sum is* -4*, such as* -1 *and* -3.

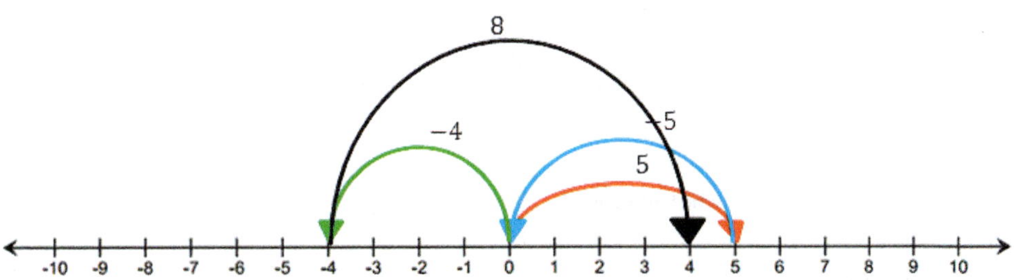

The final position is 4 units to the right of 0.

We can use smaller, curved arrows to show the number of hops or jumps that correspond to each integer. Or, we can use larger, curved arrows to show the length of the hop or jump that corresponds to the distance between the tail and the tip of the arrow along the number line. Either way, the final position is 4 units to the right of zero. Playing the Integer Game will prepare students for integer addition using arrows (vectors) in Lesson 2.

Example 3: Using the Integer Game and the Number Line

What is the sum of the card values shown? Use the counting on method on the provided number line to justify your answer.

4

5	-5	-4	8

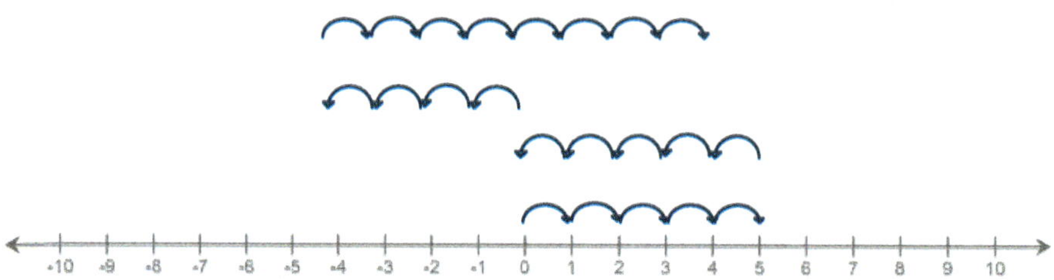

a. **What is the final position on the number line?** 4

b. **What card or combination of cards would you need to get back to 0?** -4 *or* -1 *and* -3

Exercise 2 (5 minutes): The Additive Inverse

Before students begin, the teacher highlights part of the previous example where starting at zero and counting up five units and then back down five units brings us back to zero. This is because 5 and -5 are opposites. Students work independently to answer the questions. At the end of the exercise questions, formalize the definition of *additive inverse.*

Exercise 2: The Additive Inverse

Use the number line to answer each of the following questions.

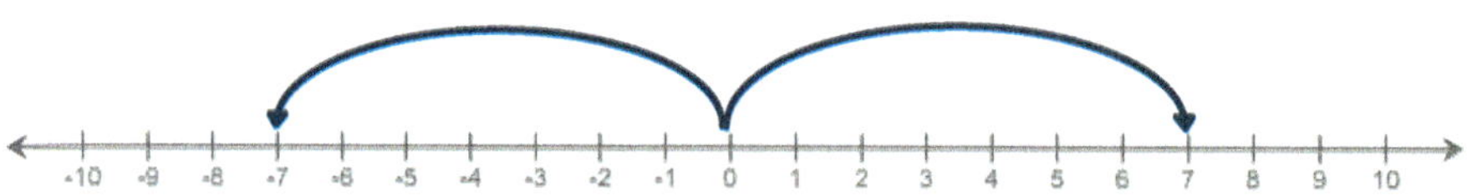

a. **How far is 7 from 0 and in which direction?** *7 units to the right*

b. **What is the opposite of 7?** -7

c. **How far is -7 from 0 and in which direction?** *7 units to the left*

d. **Thinking back to our previous work, explain how you would use the counting on method to represent the following: While playing the Integer Game, the first card selected is 7, and the second card selected is -7.**

I would start at 0 and count up 7 by moving to the right. Then, I would start counting back down from 7 to 0.

e. **What does this tell us about the sum of 7 and its opposite, -7?**

The sum of 7 and -7 equals 0.

$7 + (-7) = 0$

f. **Look at the curved arrows you drew for 7 and -7. What relationship exists between these two arrows that would support your claim about the sum of 7 and -7?**

The arrows are both the same distance from 0. They are just pointing in opposite directions.

g. **Do you think this will hold true for the sum of any number and its opposite? Why?**

I think this will be true for the sum of any number and its opposite because when you start at 0 on the number line and move in one direction, moving in the opposite direction the same number of times will always take you back to zero.

Property: For every number a, there is a number $-a$ so that $a + (-a) = 0$ and $(-a) + a = 0$.

The additive inverse of a number is a number such that the sum of the two numbers is 0. The opposite of a number satisfies this definition: For example, the opposite of 3 is -3, and $3 + (-3) = 0$. Hence, -3 is the additive inverse of 3.

The property above is usually called the existence of additive inverses.

Example 4 (5 minutes): Modeling with Real-World Examples

The purpose of this example is to introduce real-world applications of opposite quantities to make zero. The teacher holds up an Integer Game card, for example -10, to the class and models how to write a story problem.

- How would the value of this card represent a temperature?
  - *-10 could mean 10 degrees below zero.*
- How would the temperature need to change in order to get back to 0 degrees?
  - *The temperature needs to rise 10 degrees.*
- With a partner, write a story problem using money that represents the expression $200 + (-200)$.
  - *Answers will vary. Timothy has \$200 in his bank account. He owes \$200 to a friend for a bike. In terms of the loan and bank account, how much money is Timothy really worth?*

Students share their responses to the last question with the class.

Exercise 3 (10 minutes): Playing the Integer Game

> **Exercise 3: Playing the Integer Game**
>
> **Play the Integer Game with your group. Use a number line to practice counting on.**

MP.6 & MP.7

Students play the Integer Game in groups. Students practice counting using their number lines. Let students model addition on the number line. Monitor student understanding by ensuring that the direction of the arrows appropriately represents positive or negative integers.

Closing (2 minutes)

Students discuss the following questions in their groups to summarize the lesson.

- How do you model addition using a number line?
  - *When adding a positive number on a number line, you count up by moving to the right. When adding a negative number on a number line, you count down by moving to the left.*
- Using a number line, how could you find the sum of $(-5) + 6$?
  - *Start at zero, and then count down or move to the left five. From this point, count up or move to the right six.*
- Peter says he found the sum by thinking of it as $(-5) + 5 + 1$. Is this an appropriate strategy? Why do you think Peter did this?
  - *Peter did use an appropriate strategy to determine the sum of* $(-5) + 6$. *Peter did this because* 5 *and* -5 *are additive inverses, so they have a sum of zero. This made it easier to determine the sum to be* 1.
- Why is the opposite of a number also called the additive inverse? What is the sum of a number and its opposite?
  - *The opposite of a number is called the additive inverse because the two numbers' sum is zero.*

> Lesson Summary
>
> - **Add a positive number to a number by counting up from that number, and add a negative number to a number by counting down from that number.**
> - **An integer plus its opposite sum to zero.**
> - **The opposite of a number is called the additive inverse because the two numbers' sum is zero.**

Exit Ticket (5 minutes)

Name ______________________________ Date ____________

Lesson 1: Opposite Quantities Combine to Make Zero

Exit Ticket

1. Your hand starts with the 7 card. Find three different pairs that would complete your hand and result in a value of zero.

7		
7		
7		

2. Write an equation to model the sum of the situation below.

   A hydrogen atom has a zero charge because it has one negatively charged electron and one positively charged proton.

3. Write an equation for each diagram below. How are these equations alike? How are they different? What is it about the diagrams that lead to these similarities and differences?

Diagram A:

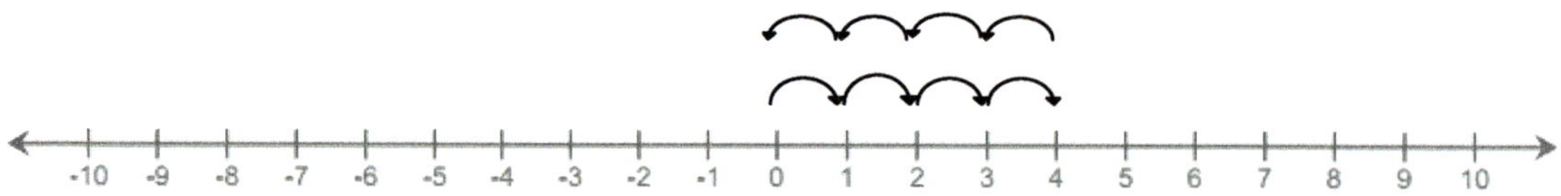

Diagram B:

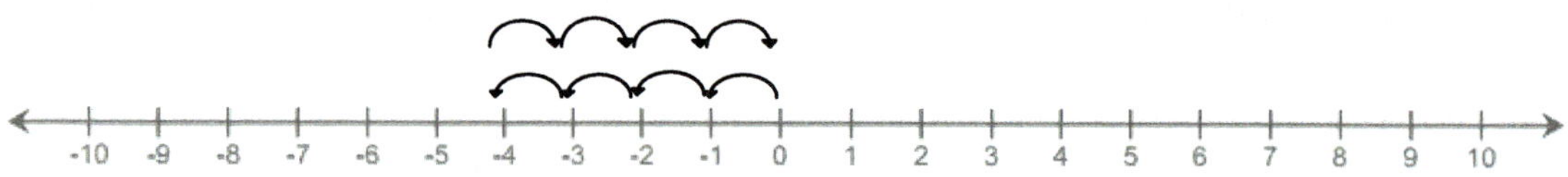

Exit Ticket Sample Solutions

1. Your hand starts with the 7 card. Find three different pairs that would complete your hand and result in a value of zero.

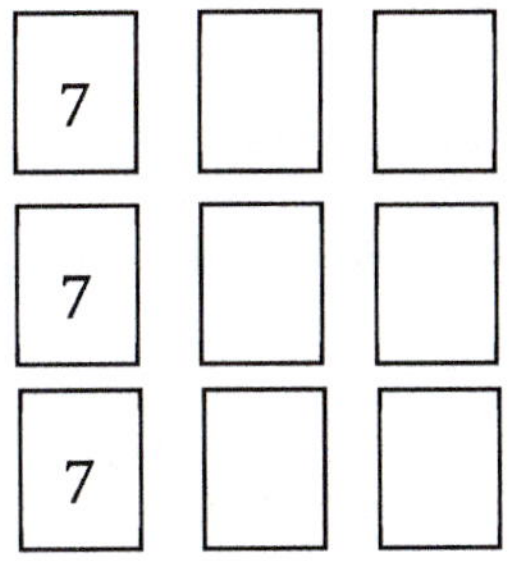

Answers will vary. $(-3$ *and* $-4)$, $(-5$ *and* $-2)$, $(-10$ *and* $3)$

2. Write an equation to model the sum of the situation below.

   A hydrogen atom has a zero charge because it has one negatively charged electron and one positively charged proton.

   $(-1) + 1 = 0$ *or* $1 + (-1) = 0$

3. Write an equation for each diagram below. How are these equations alike? How are they different? What is it about the diagrams that lead to these similarities and differences?

   Diagram A:

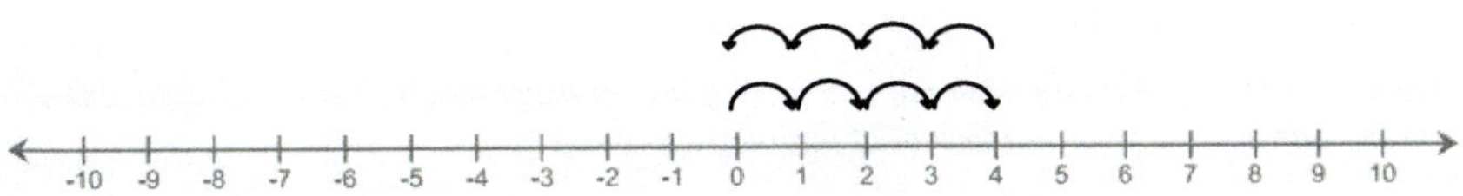

   Diagram B:

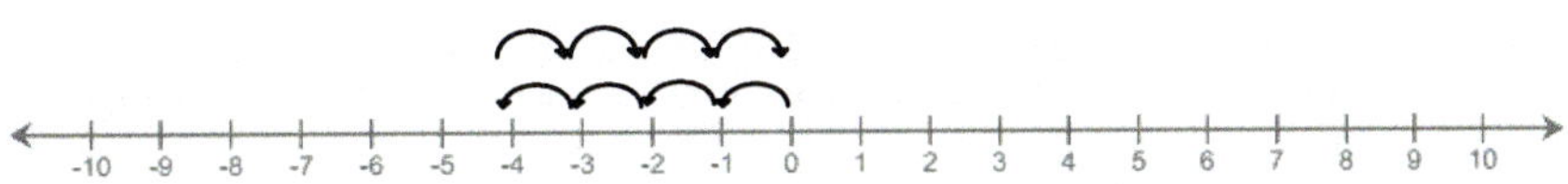

   A: $4 + (-4) = 0$

   B: $-4 + 4 = 0$

   Answers will vary. Both equations are adding 4 *and* -4*. The order of the numbers is different. The direction of A shows counting up* 4 *and then counting down* 4*. The direction of B shows counting down* 4 *and then counting up* 4*.*

   Students may also mention that both diagrams demonstrate a sum of zero, adding opposites, or adding additive inverses.

Problem Set Sample Solutions

The Problem Set provides practice with real-world situations involving the additive inverse such as temperature and money. Students also explore more scenarios from the Integer Game to provide a solid foundation for Lesson 2.

For Problems 1 and 2, refer to the Integer Game.

1. **You have two cards with a sum of (-12) in your hand.**

   a. **What two cards could you have?**

   Answers will vary. $(-6$ *and* $-6)$

   b. **You add two more cards to your hand, but the total sum of the cards remains the same, (-12). Give some different examples of two cards you could choose.**

   Answers will vary, but numbers must be opposites. $(-2$ *and* $2)$ *and* $(4$ *and* $-4)$

2. **Choose one card value and its additive inverse. Choose from the list below to write a real-world story problem that would model their sum.**

   a. **Elevation: above and below sea level**

   Answers will vary. (A scuba diver is 20 *feet below sea level. He had to rise* 20 *feet in order to get back on the boat.)*

   b. **Money: credits and debits, deposits and withdrawals**

   Answers will vary. (The bank charges a fee of $\$5$ *for replacing a lost debit card. If you make a deposit of* $\$5$*, what would be the sum of the fee and the deposit?)*

   c. **Temperature: above and below 0 degrees**

   Answers will vary. (The temperature of one room is 5 *degrees above* 0*. The temperature of another room is* 5 *degrees below zero. What is the sum of both temperatures?)*

   d. **Football: loss and gain of yards**

   Answers will vary. (A football player gained 25 *yards on the first play. On the second play, he lost* 25 *yards. What is his net yardage after both plays?)*

3. **On the number line below, the numbers h and k are the same distance from 0. Write an equation to express the value of $h + k$. Explain.**

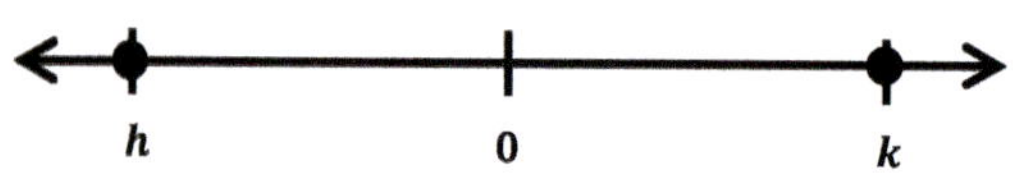

$h + k = 0$ because their absolute values are equal, but their directions are opposite. k is the additive inverse of h, and h is the additive inverse of k because they are the same distance from 0. Therefore, the sum of k and h is 0, because additive inverses have a sum of 0.

4. During a football game, Kevin gained five yards on the first play. Then he lost seven yards on the second play. How many yards does Kevin need on the next play to get the team back to where they were when they started? Show your work.

He has to gain 2 *yards.*

$5 + (-7) + 2 = 0$, $5 + (-7) = -2$, *and* $-2 + 2 = 0$.

5. Write an addition number sentence that corresponds to the arrows below.

$10 + (-5) + (-5) = 0$

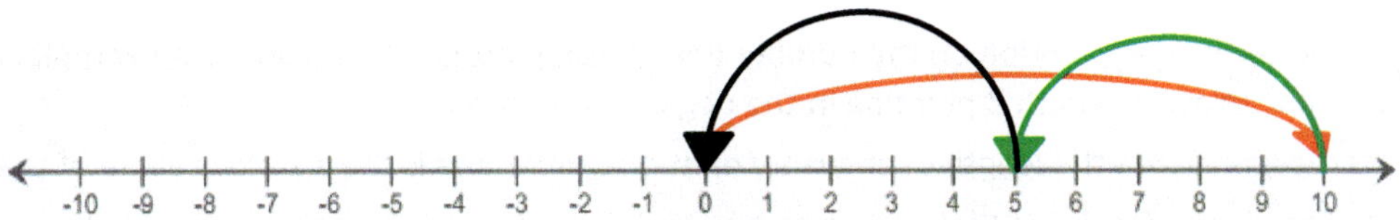

Lesson 2: Using the Number Line to Model the Addition of Integers

Student Outcomes

- Students model integer addition on the number line by using horizontal arrows; for example, an arrow for -2 is a horizontal arrow of length 2 pointing in the negative direction.
- Students recognize that the length of an arrow on the number line is the absolute value of the integer.
- Students add arrows (realizing that adding arrows is the same as combining numbers in the Integer Game). Given several arrows, students indicate the number that the arrows represent (the sum).

Classwork

Exercise 1 (5 minutes): Real-World Introduction to Integer Addition

Scaffolding:

Create an anchor poster for the additive inverse to help access prior knowledge of number line features including arrow placement and direction and ordering of positive and negative numbers.

MP.4 Students answer the following question independently as the teacher circulates around the room providing guidance and feedback as needed. Students focus on how to represent the answer using both an equation and a number line diagram and make the connection between both representations.

Exercise 1: Real-World Introduction to Integer Addition

Answer the questions below.

a. **Suppose you received $10 from your grandmother for your birthday. You spent $4 on snacks. Using addition, how would you write an equation to represent this situation?**

$10 + (-4) = 6$

b. **How would you model your equation on a number line to show your answer?**

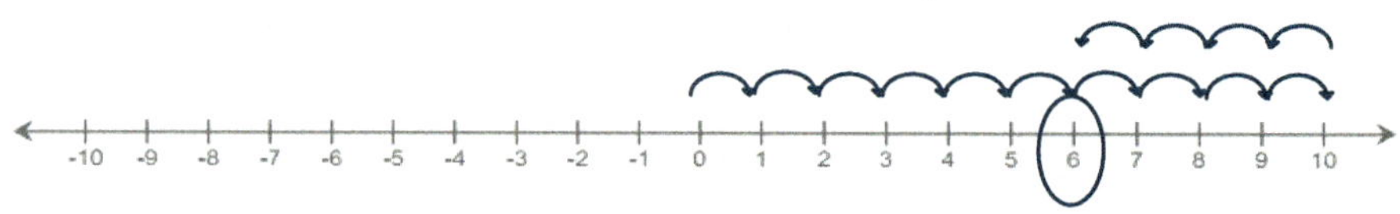

Real-world situations can be modeled with equations and represented on a real number line. In this exercise, positive ten represents the $10 given as a birthday gift because it is a gain. Negative four represents the $4 spent on snacks because it is a loss. Gaining $10 and then taking away $4 leaves you with $6.

Example 1 (5 minutes): Modeling Addition on the Number Line

The teacher models addition on a number line using straight arrows (vectors) to find the sum of $-2 + 3$. Elicit student responses to assist in creating the steps. Students record the steps and diagram.

- Place the tail of the arrow on 0.
- Draw the arrow 2 units to the left of 0, and stop at -2. The direction of the arrow is to the left since you are counting down from 0.
- Start the next arrow at the end of the first arrow, or at -2.
- Draw the second arrow 3 units to the right since you are counting up from -2.
- Stop at 1.
- Circle the number at which the second arrow ends to indicate the ending value.

> *Scaffolding:*
> - Use counters or chips to transfer prior learning of additive inverse or zero pairs.
> - Create a number line model on the floor for kinesthetic and visual learners.

Using the example, model a real-world story problem for the class.

- If the temperature outside were 2 degrees below zero, and it increased by 3 degrees, the new temperature outside would be 1 degree.

Have students share a story problem involving temperature, money, or sea level that would describe the number line model. Select a few students to share their answers with the class.

  - *Answers will vary. I owed my brother* $\$2$*, and my dad gave me* $\$3$*. I paid my brother, and now I have* $\$1$ *left over.*

Example 1: Modeling Addition on the Number Line

Complete the steps to find the sum of $-2 + 3$ by filling in the blanks. Model the equation using straight arrows called *vectors* on the number line below.

a. **Place the tail of the arrow on** __0__.

b. **Draw the arrow 2 units to the left of 0, and stop at** __-2__. **The direction of the arrow is to the** __*left*__ **since you are counting down from 0.**

c. **Start the next arrow at the end of the first arrow, or at** __-2__.

d. **Draw the second arrow** __3__ **units to the right since you are counting up from -2.**

e. **Stop at** __1__.

f. **Circle the number at which the second arrow ends to indicate the ending value.**

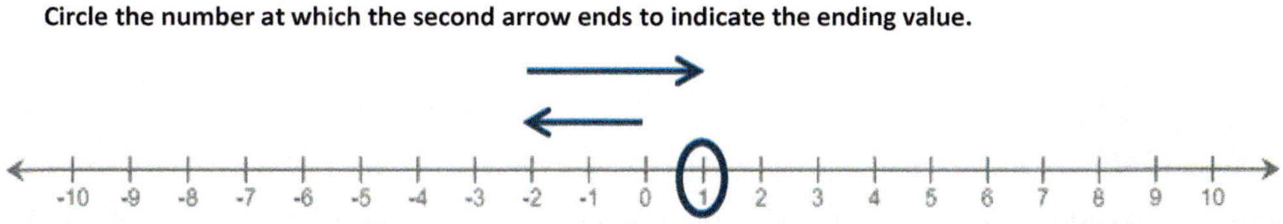

g. **Repeat the process from parts (a)–(f) for the expression $3 + (-2)$.**

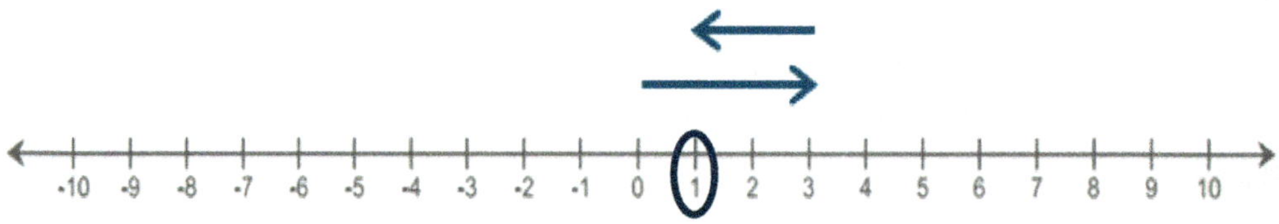

h. **What can you say about the sum of $-2 + 3$ and $3 + (-2)$? Does order matter when adding numbers? Why or why not?**

$-2 + 3$ is the same as $3 + (-2)$ because they both equal 1. The order does not matter when adding numbers because addition is commutative.

Example 2 (3 minutes): Expressing Absolute Value as the Length of an Arrow on the Real Number Line

The teacher models absolute value as the length of an arrow. Students recall that absolute value represents distance.

Example 2: Expressing Absolute Value as the Length of an Arrow on the Real Number Line

a. **How does absolute value determine the arrow length for -2? Use the number line provided to support your answer.**

$|-2| = 2$, so the arrow is 2 units long. Because -2 is a negative number, the arrow points to the left.

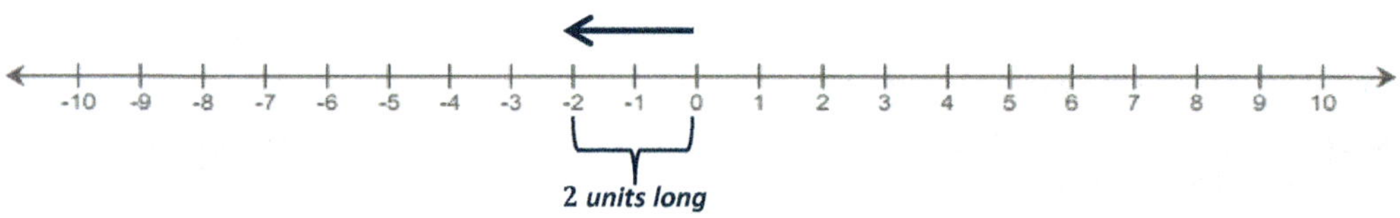

b. **How does the absolute value determine the arrow length for 3? Use the number line provided to support your answer.**

$|3| = 3$, so the arrow is 3 units long. Because 3 is positive, the arrow points to the right.

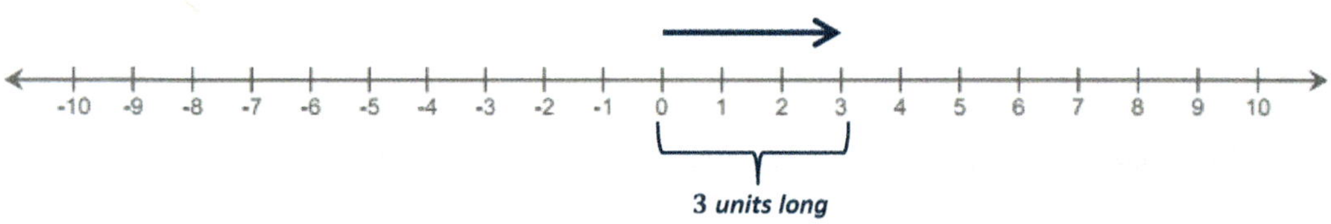

c. **Describe how the absolute value helps you represent -10 on a number line.**

The absolute value can help me because it tells me how long my arrow should be when starting at 0 on the real number line. The $|-10| = 10$, so my arrow will be 10 units in length.

Exercise 2 (5 minutes)

MP.6

Students work independently to create a number line model to represent each of the expressions below. After 2–3 minutes, students are selected to share their responses and work with the class. Monitor student work by paying careful attention to common mistakes such as miscounting, not lining up arrows head-to-tail, and starting both arrows at 0.

Exercise 2

Create a number line model to represent each of the expressions below.

a. $-6 + 4$

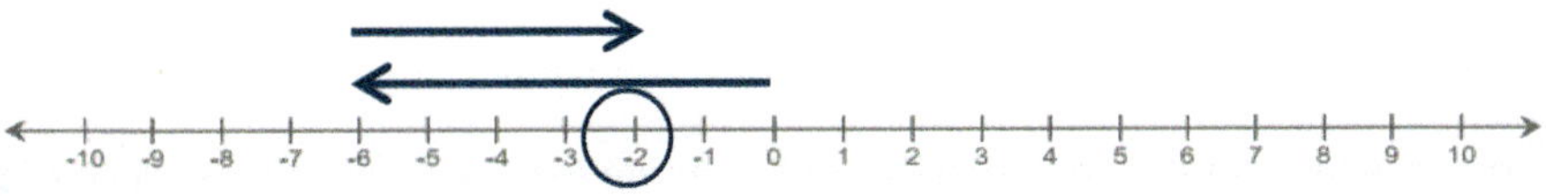

b. $3 + (-8)$

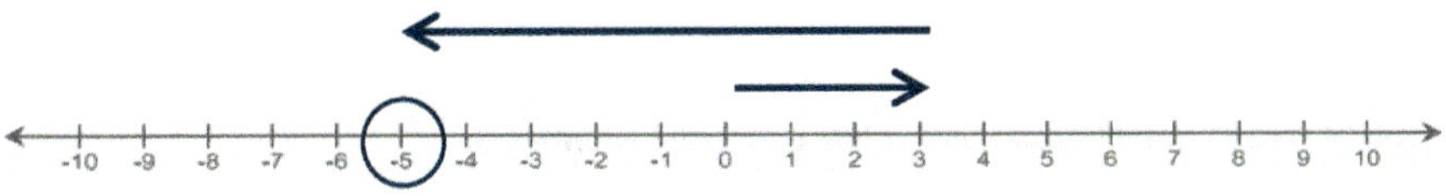

Scaffolding:

Have early finishers explain how absolute value determined the arrow lengths for each of the addends and how they knew each arrow's direction.

Example 3 (5 minutes): Finding Sums on a Real Number Line Model

The teacher refers to the Integer Game from Lesson 1. Pose discussion questions to the class.

Example 3: Finding Sums on a Real Number Line Model

Find the sum of the integers represented in the diagram below.

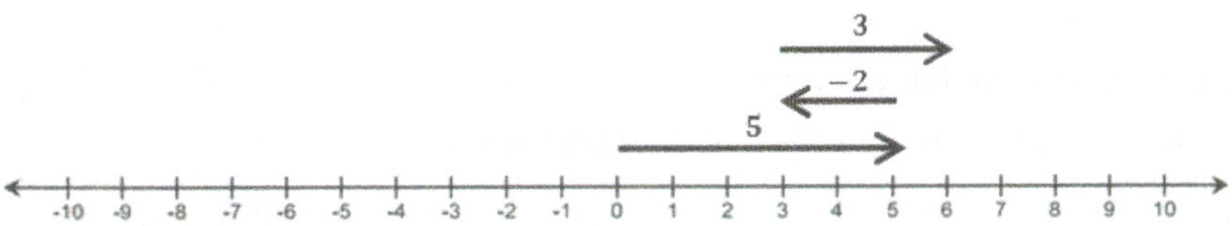

a. **Write an equation to express the sum.**

$5 + (-2) + 3 = 6$

b. **What three cards are represented in this model? How did you know?**

The cards are 5, -2, *and* 3 *because the arrows show their lengths.*

c. **In what ways does this model differ from the ones we used in Lesson 1?**

In Lesson 1, a movement of 5 *units was shown with* 5 *separate hops. In this lesson,* 5 *units are shown as one total movement with a straight arrow. Both represent the same total movement.*

d. **Can you make a connection between the sum of 6 and where the third arrow ends on the number line?**

The final position of the third arrow is 6. *This means that the sum is* 6.

e. **Would the sum change if we changed the order in which we add the numbers, for example, $(-2) + 3 + 5$?**

No, because addition is commutative. Order does not matter.

Scaffolding:

- Have students use their same cards to create a different addition number sentence and a new number line representation.
- Have students examine how the diagram changes when the order of addition changes to reinforce the commutative property.

f. **Would the diagram change? If so, how?**

Yes, the first arrow would start at 0 and point left 2 units. The second arrow would start at -2 and point right 3 units. The third arrow would start at 1 and point 5 units right but still end on 6.

Exercise 3 (14 minutes)

In groups of 3–4, students play the Integer Game[1]. The objective of the game for Lesson 2 is to get as close to 0 as possible. During play, students work independently to create an equation and number line diagram to model integer addition. Monitor the classroom and ask probing questions.

Exercise 3

Play the Integer Game with your group. Use a number line to practice counting on.

Closing (3 minutes)

The teacher initiates whole-group discussion, prompting students to verbally state the answers to the following questions:

- How can we use a number line to model and find the sum of $-8 + 5$?
  - *We would start at 0 and then draw an arrow eight units to the left to represent -8. From the end of this arrow, we would draw an arrow five units to the right to represent 5. The number the final arrow ends on is the sum of $-8 + 5$.*
- What does the absolute value of a number tell you?
  - *The absolute value of a number tells us the length of the arrow.*

Lesson Summary

- **On a number line, arrows are used to represent integers; they show length and direction.**
- **The length of an arrow on the number line is the absolute value of the integer.**
- **Adding several arrows is the same as combining integers in the Integer Game.**
- **The sum of several arrows is the final position of the last arrow.**

Exit Ticket (5 minutes)

[1]Refer to the Integer Game outline for player rules.

Name ______________________________ Date ______________

Lesson 2: Using the Number Line to Model the Addition of Integers

Exit Ticket

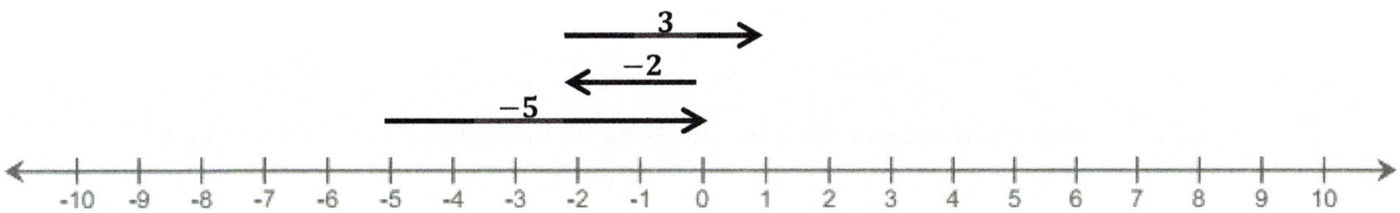

Jessica made the addition model below of the expression $(-5) + (-2) + 3$.

a. Do the arrows correctly represent the numbers that Jessica is using in her expression?

b. Jessica used the number line diagram above to conclude that the sum of the three numbers is 1. Is she correct?

c. If she is incorrect, find the sum, and draw the correct model.

d. Write a real-world situation that would represent the sum.

Exit Ticket Sample Solutions

Jessica made the addition model below of the expression $(-5) + (-2) + 3$.

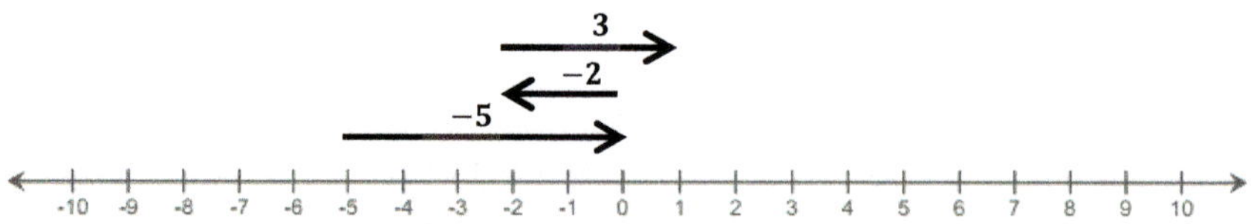

a. **Do the arrows correctly represent the numbers that Jessica is using in her expression?**

No. Jessica started her first arrow at -5 instead of 0. Negative numbers should be shown as counting down, so the arrow should have started at 0 and pointed left, ending on -5. The other arrows are drawn correctly, but they are in the wrong places because the starting arrow is in the wrong place.

b. **Jessica used the number line diagram above to conclude that the sum of the three numbers is 1. Is she correct?**

Jessica is incorrect.

c. **If she is incorrect, find the sum, and draw the correct model.**

The sum should be -4. $-5 + (-2) + 3 = -4$.

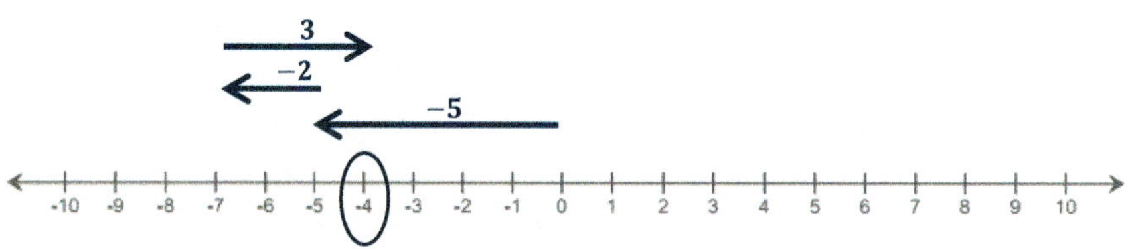

d. **Write a real-world situation that would represent the sum.**

Answers will vary. A football team lost 5 yards on the first play. On the second play, the team lost another 2 yards. Then, the team gained 3 yards. After three plays, the team had a total yardage of -4 yards.

Problem Set Sample Solutions

The Problem Set provides students practice with integer addition using the Integer Game, number lines, and story problems. Students should show their work with accuracy in order to demonstrate mastery.

Represent Problems 1–3 using both a number line diagram and an equation.

1. **David and Victoria are playing the Integer Card Game. David drew three cards, -6, 12, and -4. What is the sum of the cards in his hand? Model your answer on the number line below.**

   $(-6) + 12 + (-4) = 2$

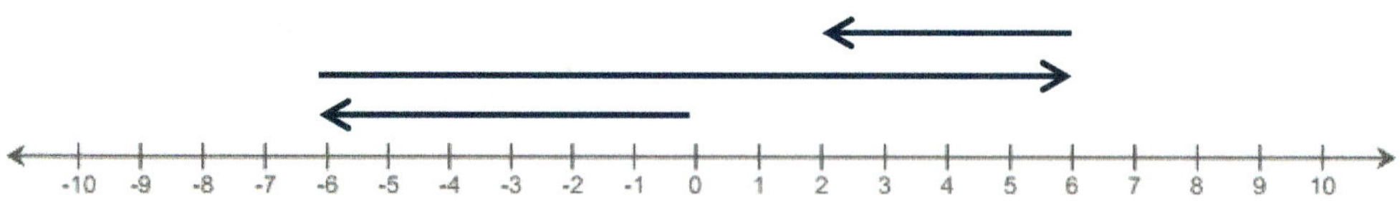

2. In the Integer Card Game, you drew the cards, 2, 8, and -11. Your partner gave you a 7 from his hand.

   a. What is your total? Model your answer on the number line below.

   $2 + 8 + (-11) + 7 = 6$

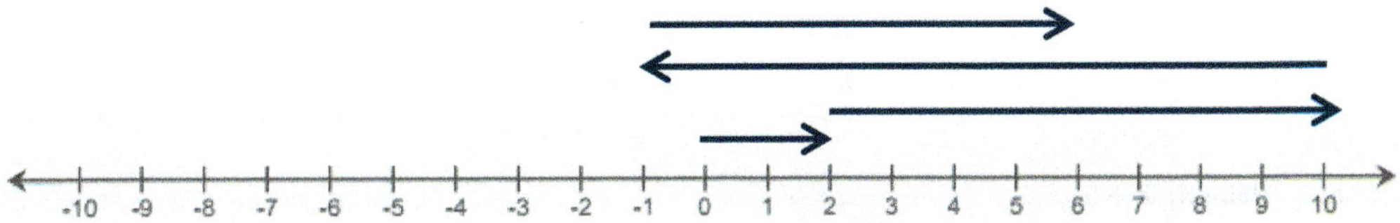

   b. What card(s) would you need to get your score back to zero? Explain. Use and explain the term *additive inverse* in your answer.

   You would need any combination of cards that sum to -6 *because the additive inverse of* 6 *is* -6.

   $6 + (-6) = 0$

3. If a football player gains 40 yards on a play, but on the next play, he loses 10 yards, what would his total yards be for the game if he ran for another 60 yards? What did you count by to label the units on your number line?

   90 *yards because* $40 + (-10) + 60 = 90$. *Student answers may vary, but they should not choose to count by* 1.

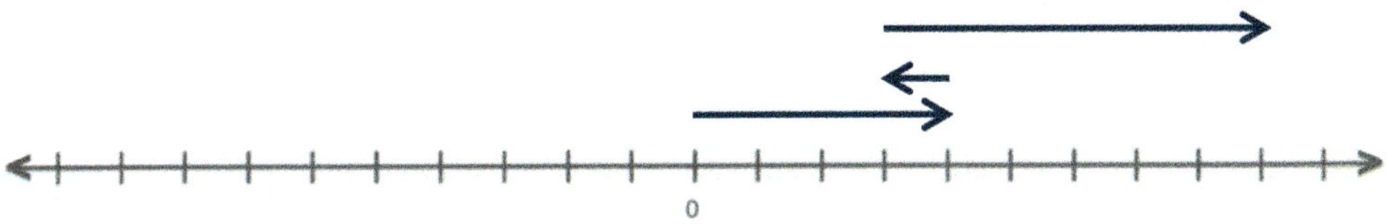

4. Find the sums.

   a. $-2 + 9$

   7

   b. $-8 + -8$

   -16

   c. $-4 + (-6) + 10$

   0

   d. $5 + 7 + (-11)$

   1

5. Mark an integer between 1 and 5 on a number line, and label it point Z. Then, locate and label each of the following points by finding the sums.

Answers will vary. Sample student response below.

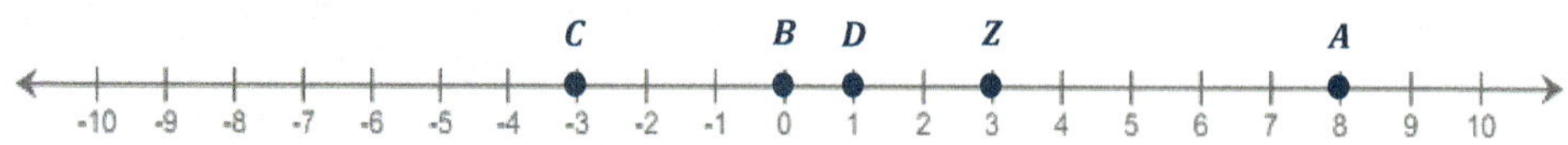

a. Point A: $Z + 5$

Point A: $3 + 5 = 8$

b. Point B: $Z + (-3)$

Point B: $3 + (-3) = 0$

c. Point C: $(-4) + (-2) + Z$

Point C: $(-4) + (-2) + 3 = -3$

d. Point D: $-3 + Z + 1$

Point D: $-3 + 3 + 1 = 1$

6. Write a story problem that would model the sum of the arrows in the number diagram below.

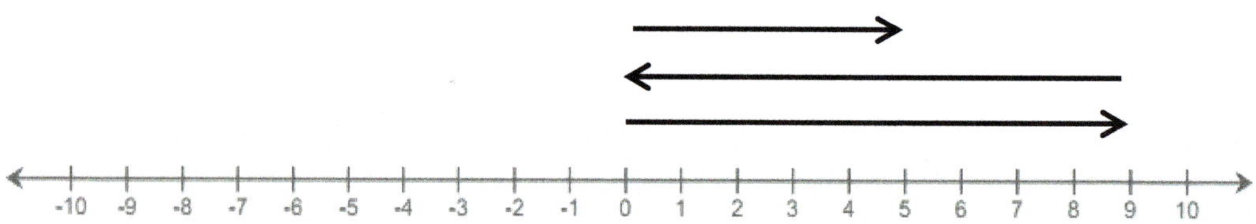

Answers will vary. Jill got on an elevator and went to the 9th floor. She accidently pressed the down button and went back to the lobby. She pressed the button for the 5th floor and got off the elevator.

7. Do the arrows correctly represent the equation $4 + (-7) + 5 = 2$? If not, draw a correct model below.

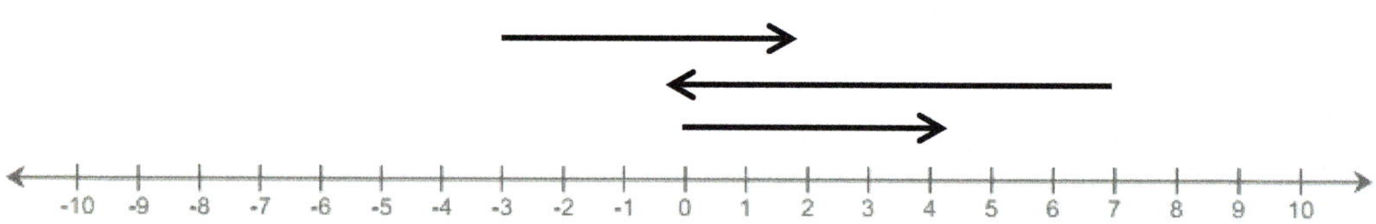

No, the arrows are incorrect. The correct model is shown.

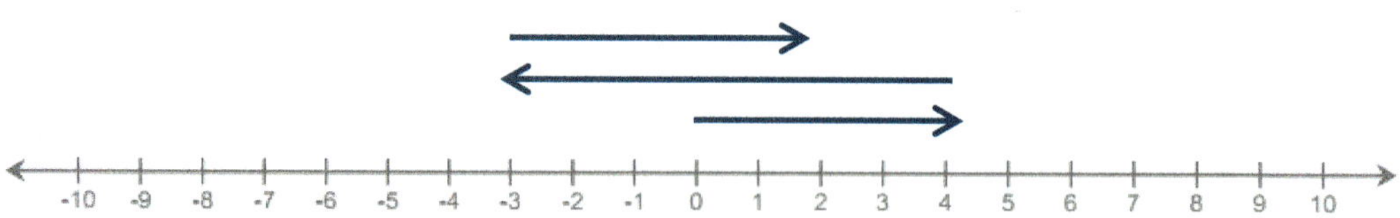

Lesson 3: Understanding Addition of Integers

Student Outcomes

- Students understand addition of integers as putting together or counting up. For negative numbers "counting up" is actually counting down.
- Students use arrows to show the sum of two integers, $p + q$, on a number line and to show that the sum is distance $|q|$ from p to the right if q is positive and to the left if q is negative.
- Students refer back to the Integer Game to reinforce their understanding of addition.

Lesson Notes

We have addressed the letters p and q in this lesson to help show the relationship of this lesson to the standard **7.NS.A1b**. It is not important to use p and q (or p-value and q-value) in your lesson.

Classwork

Exercise 1 (15 minutes): Addition Using the Integer Game

Exercise 1: Addition Using the Integer Game

Play the Integer Game with your group without using a number line.

Scaffolding:

Allow for the use of a number line for English language learners if needed.

In pairs, students play a modified version of the Integer Game[1] without a number line. Monitor student play and ask probing questions. When students share at the end of the game, see if anyone used the concept of additive inverse, if the opportunity occurred, when adding.

[1]Refer to the Integer Game outline for complete player rules. In Exercise 1, cards are shuffled and placed facedown. Players draw three cards each and calculate the sums of their hands. Once they each have the sum of their three cards, players put down their cards face up. Next, they find the sum of all six cards that they have collectively.

Example 1 (10 minutes): Counting On to Express the Sum as Absolute Value on a Number Line

The teacher leads whole-class instruction using vector addition to (1) review the sum of two integers on a real number line horizontally and vertically and (2) show that the sum is the distance of the absolute value of the q-value (second addend) from the p-value (first addend).

Example 1: Counting On to Express the Sum as Absolute Value on a Number Line

Model of Counting Up

$$2 + 4 = 6$$

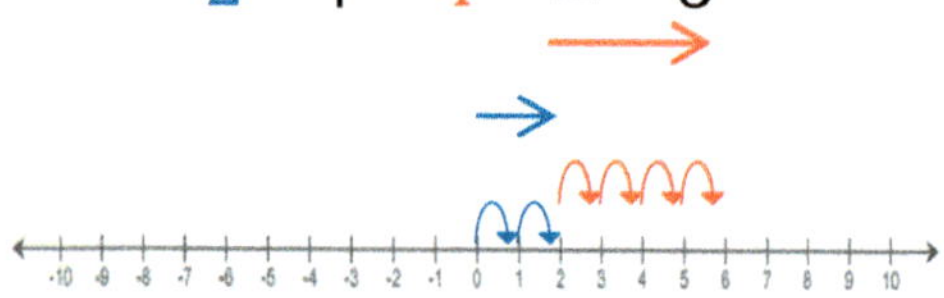

Model of Counting Down

$$2 + (-4) = -2$$

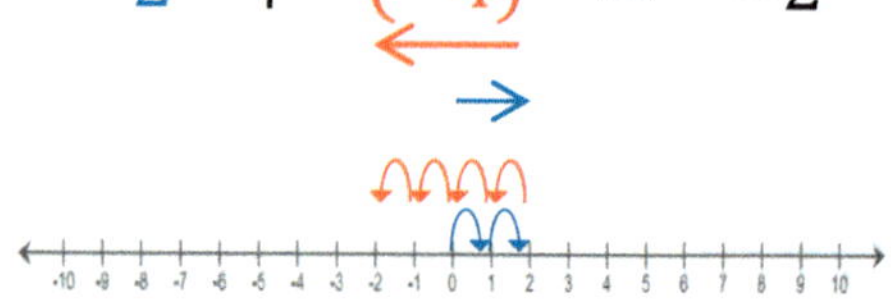

Horizontal Number Line Model of Counting Up

$$2 + 4 = 6$$

p-value
Start at 0; count up 2 units to the right.
Arrow is 2 units long and points to the right.

q-value
$|4| = 4$
Absolute value of 4 is 4.
Start at 2; count up 4 units to the right.
Arrow is 4 units long and points to the right.

Sum
6 is 4 units away from 2.

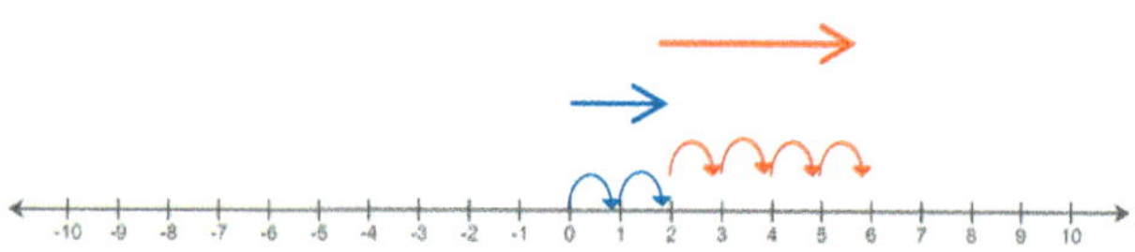

Horizontal Number Line Model of Counting Down

$$2 + (-4) = -2$$

p-value
Start at 0; count up 2 units to the right.
Arrow is 2 units long and points to the right.

q-value
$|-4| = 4$
Absolute value of -4 is 4.
Start at 2; count down 4 units to the left.
Arrow is 4 units long and points to the left.

Sum
-2 is 4 units away from 2.

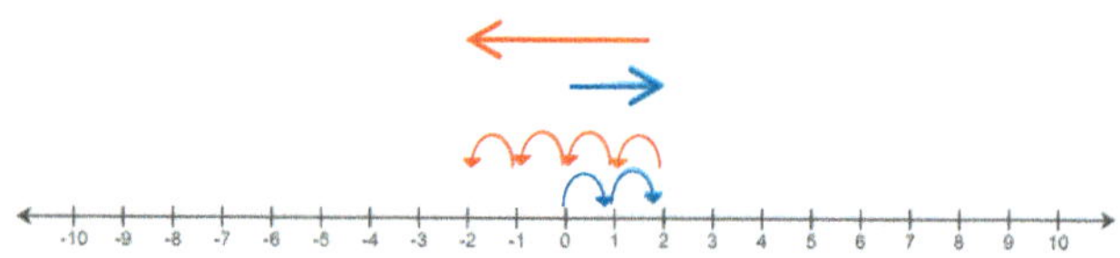

Vertical Number Line Model of Counting Up

$2 + 4 = 6$

p-value
Start at 0; count up 2 units above 0. Arrow is 2 units long and points up.

q-value
$|4| = 4$
Absolute value of 4 is 4.
Start at 2; count up 4 units.
Arrow is 4 units long and points up.

Sum
6 is 4 units above 2.

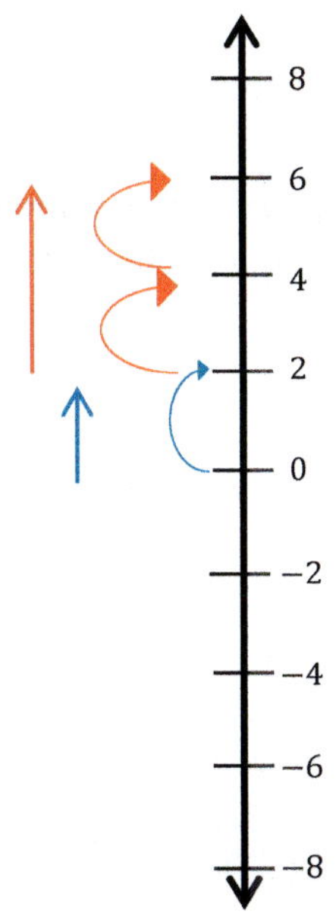

Vertical Number Line Model of Counting Down

$2 + (-4) = -2$

p-value
Start at 0; count up 2 units above 0. Arrow is 2 units long and points up.

q-value
$|-4| = 4$
Absolute value of -4 is 4.
Start at 2; count down 4 units.
Arrow is 4 units long and points down.

Sum
-2 is 4 units below 2.

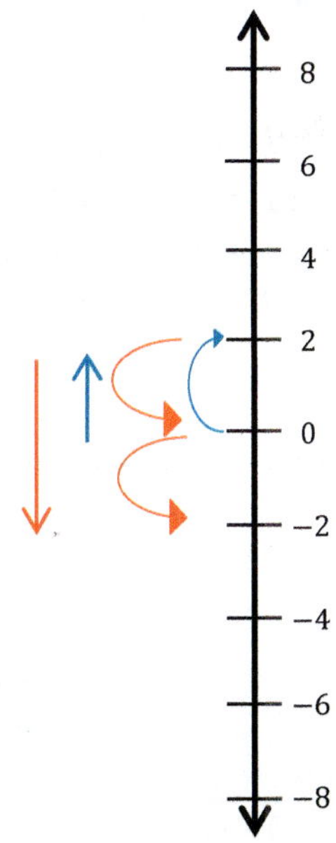

The teacher poses the following questions to the class for open discussion. Students record their responses in the space provided.

Counting up -4 is the same as *the opposite of counting up* 4 and also means counting down 4.

a. **For each example above, what is the distance between 2 and the sum?**

4 units

b. **Does the sum lie to the right or left of 2 on a horizontal number line? Above or below on a vertical number line?**

Horizontal: On the first model, the sum lies to the right of 2. On the second model, it lies to the left of 2.

Vertical: On the first model, the sum lies above 2. On the second model, it lies below 2.

c. **Given the expression $54 + 81$, determine, without finding the sum, the distance between 54 and the sum. Explain.**

The distance will be 81 units. When the q-value is positive, the sum will be to the right of (or above) the p-value the same number of units as the q-value.

d. **Is the sum to the right or left of 54 on the horizontal number line? Above or below on a vertical number line?**

The sum is to the right of 54 on a horizontal number line and above 54 on a vertical number line.

e. **Given the expression $14 + (-3)$, determine, without finding the sum, the distance between 14 and the sum. Explain.**

The distance will be 3 units. When the q-value is negative, the sum will be to the left of (or below) the p-value the same number of units as the q-value.

f. **Is the sum to the right or left of 14 on the number line? Above or below on a vertical number line?**

The sum is to the left of 14 on a horizontal number line and below 14 on a vertical number line.

Scaffolding:

- Review the concept of *sum* with the whole class for English language learners.
- Provide written stems for English language learners. For example, "The sum is ____ units to the ____ of ____."

Exercise 2 (5 minutes)

Students work in pairs to create a number line model to represent each of the following expressions. After 2 or 3 minutes, students are selected to share their responses and work with the class. Ask students to describe the sum using distance from the first addend along the number line.

Exercise 2

Work with a partner to create a horizontal number line model to represent each of the following expressions. What is the sum?

a. $-5 + 3$

$-5 + 3 = -2$. *The sum is 3 units to the right of -5.*

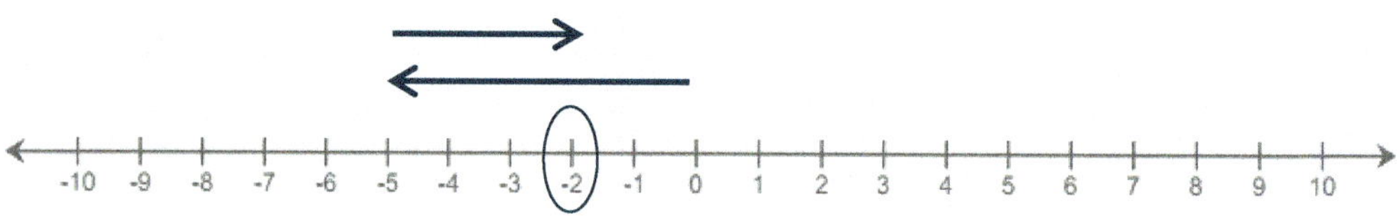

b. $-6 + (-2)$

$-6 + (-2) = -8$. *The sum is 2 units to the left of -6.*

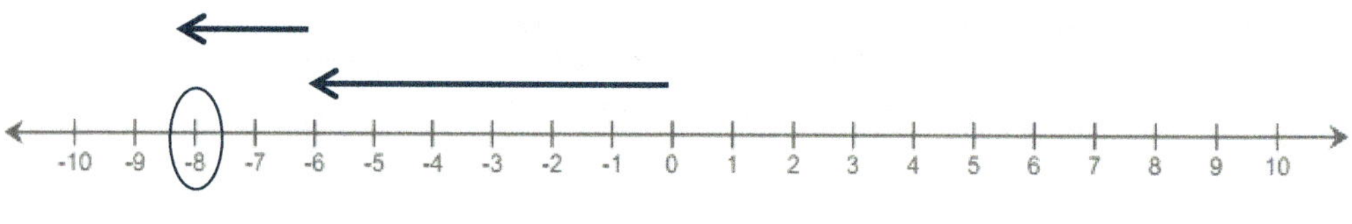

c. $7 + (-8)$

$7 + (-8) = -1$. *The sum is 8 units to the left of 7.*

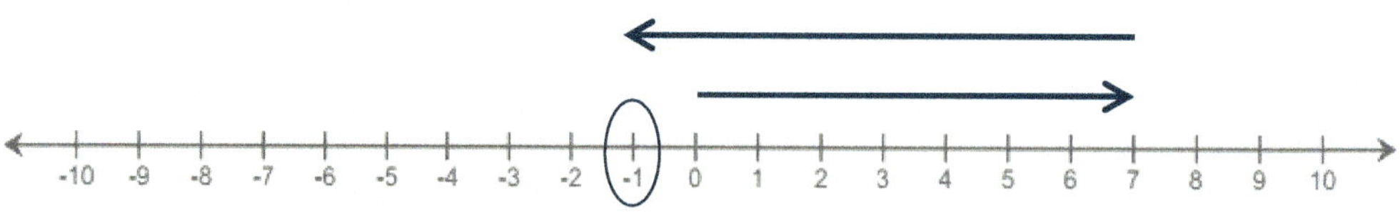

Exercise 3 (5 minutes): Writing an Equation Using Verbal Descriptions

Students continue to work in pairs to complete the following task.

Exercise 3: Writing an Equation Using Verbal Descriptions

Write an equation, and using the number line, create an *arrow diagram* given the following information:

The sum of 6 and a number is 15 units to the left of 6 on the number line.

Equation:

$6 + (-15) = -9$

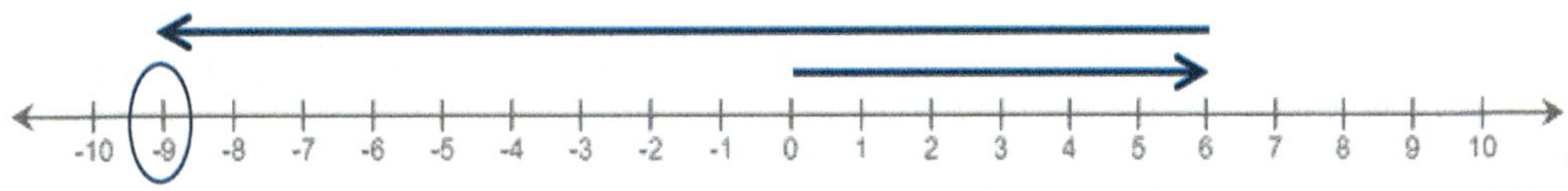

Closing (3 minutes)

The teacher uses whole-group discussion with students verbally stating the answers to the following questions.

- What role does the $|-16| = 16$ play in modeling the expression $2 + (-16)$?
  - *The absolute value of the second value (q) represents the distance between the first addend (p) and the sum.*
- What is one important fact to remember when modeling addition on a horizontal number line? On a vertical number line?
- *One important fact to remember when adding integers on a number line is that counting up a negative number of times is the same as counting down.* What is the difference between counting up and counting down?
  - *Counting up represents a positive addend, and counting down represents a negative addend.*

Lesson Summary

- **Adding an integer to a number can be represented on a number line as counting up when the integer is positive (just like whole numbers) and counting down when the integer is negative.**
- **Arrows can be used to represent the sum of two integers on a number line.**

Exit Ticket (7 minutes)

Name ____ Date ________________

Lesson 3: Understanding Addition of Integers

Exit Ticket

1. Refer to the diagram to the right.

   a. Write an equation for the diagram to the right. ____________________

   b. Find the sum. ____________________

   c. Describe the sum in terms of the distance from the first addend. Explain.

   d. What integers do the arrows represent? ____________________

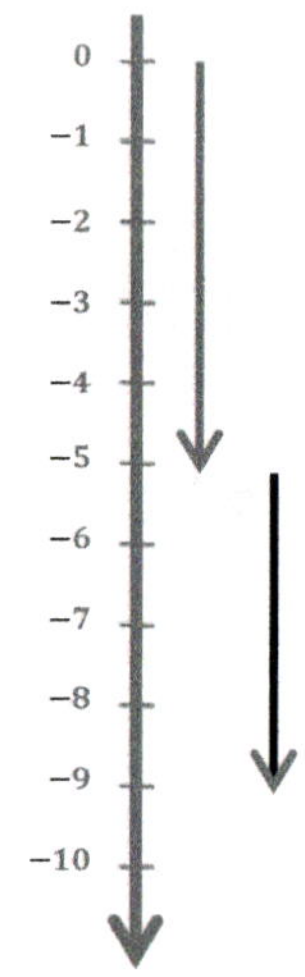

2. Jenna and Jay are playing the Integer Game. Below are the two cards they selected.

   a. How do the models for these two addition problems differ on a number line? How are they the same?

Jenna's Hand

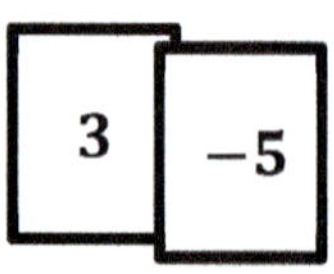

Jay's Hand

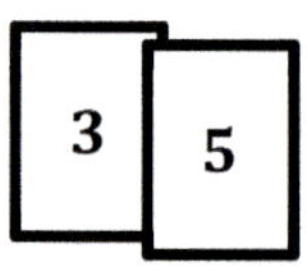

   b. If the order of the cards changed, how do the models for these two addition problems differ on a number line? How are they the same?

Jenna's Hand

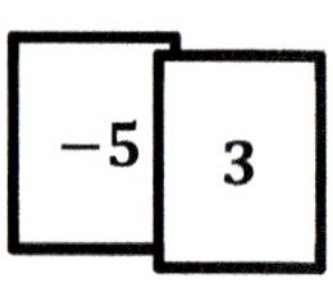

Jay's Hand

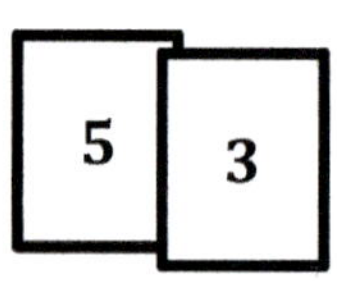

Exit Ticket Sample Solutions

1. Refer to the diagram to the right.

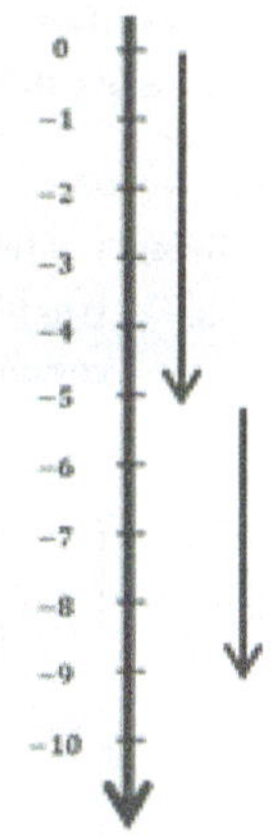

   a. Write an equation for the diagram below.

      $-5 + (-4) = -9$

   b. Find the sum.

      -9

   c. Describe the sum in terms of the distance from the first addend. Explain.

      The sum is 4 units below -5 because $|-4| = 4$. I counted down from -5 four times and stopped at -9.

   d. What integers do the arrows represent?

      The arrows represent the integers -4 and -5.

2. Jenna and Jay are playing the Integer Game. Below are the two cards they selected.

   a. How do the models for these two addition problems differ on a number line? How are they the same?

      Jenna's Hand

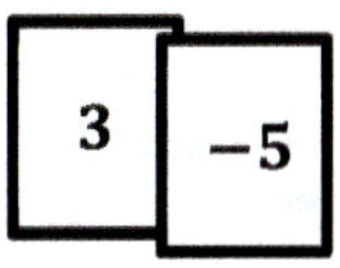

      Jay's Hand

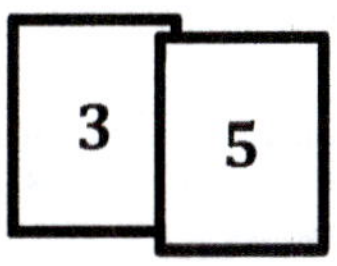

      The p-values are the same. They are both 3, so the heads of the first arrows will be at the same point on the number line. The sums will both be five units from this point but in opposite directions.

   b. If the order of the cards changed, how do the models for these two addition problems differ on a number line? How are they the same?

      Jenna's Hand

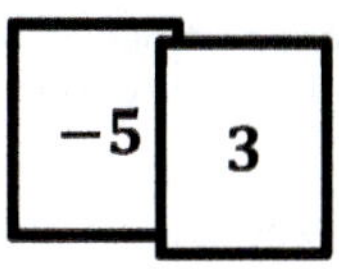

      Jay's Hand

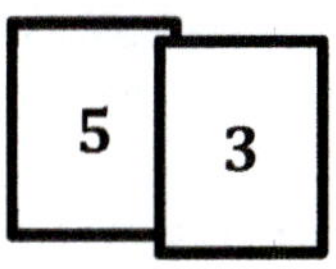

      The first addends are different, so the head of the first arrow in each model will be at different points on the number line. The sums are both three units to the right of the first addend.

Problem Set Sample Solutions

Practice problems help students build fluency and improve accuracy when adding integers with and without the use of a number line. Students need to be comfortable with using vectors to represent integers on the number line, including the application of absolute value to represent the length of a vector.

1. Below is a table showing the change in temperature from morning to afternoon for one week.

   a. Use the vertical number line to help you complete the table. As an example, the first row is completed for you.

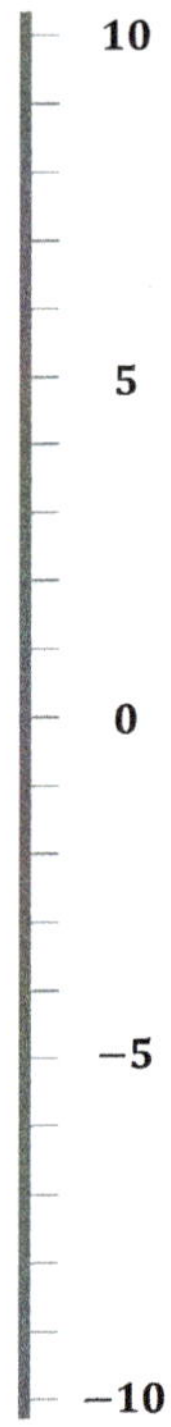

Change in Temperatures from Morning to Afternoon

Morning Temperature	Change	Afternoon Temperature	Equation
$1°C$	Rise of $3°C$	$4°C$	$1+3=4$
$2°C$	Rise of $8°C$	$10°C$	$2+8=10$
$-2°C$	Fall of $6°C$	$-8°C$	$-2+(-6)=-8$
$-4°C$	Rise of $7°C$	$3°C$	$-4+7=3$
$6°C$	Fall of $9°C$	$-3°C$	$6+(-9)=-3$
$-5°C$	Fall of $5°C$	$-10°C$	$-5+(-5)=-10$
$7°C$	Fall of $7°C$	$0°C$	$7+(-7)=0$

   b. Do you agree or disagree with the following statement: "A rise of $-7°C$" means "a fall of $7°C$?" Explain. (Note: No one would ever say, "A rise of -7 degrees"; however, mathematically speaking, it is an equivalent phrase.)

   Sample response: I agree with this statement because a rise of -7 is the opposite of a rise of 7. The opposite of a rise of 7 is a fall of 7.

For Problems 2–3, refer to the Integer Game.

2. Terry selected two cards. The sum of her cards is -10.

   a. Can both cards be positive? Explain why or why not.

   No. In order for the sum to be -10, at least one of the addends would have to be negative. If both cards are positive, then Terry would count up twice going to the right. Negative integers are to the left of 0.

   b. Can one of the cards be positive and the other be negative? Explain why or why not.

   Yes. Since both cards cannot be positive, this means that one can be positive and the other negative. She could have -11 and 1 or -12 and 2. The card with the greatest absolute value would have to be negative.

   c. Can both cards be negative? Explain why or why not.

   Yes, both cards could be negative. She could have -8 and -2. On a number line, the sum of two negative integers will be to the left of 0.

3. When playing the Integer Game, the first two cards you selected were -8 and -10.

   a. What is the value of your hand? Write an equation to justify your answer.

   $-8 + (-10) = -18$

   b. For part (a), what is the distance of the sum from -8? Does the sum lie to the right or left of -8 on the number line?

   The distance is 10 *units from* -8*, and it lies to the left of* -8 *on the number line.*

   c. If you discarded the -10 and then selected a 10, what would be the value of your hand? Write an equation to justify your answer.

   The value of the hand would be 2. $-8 + 10 = 2$.

4. Given the expression $67 + (-35)$, can you determine, without finding the sum, the distance between 67 and the sum? Is the sum to the right or left of 67 on the number line?

   The distance would be 35 *units from* 67. *The sum is to the left of* 67 *on the number line.*

5. Use the information given below to write an equation. Then create an *arrow diagram* of this equation on the number line provided below.

   The sum of -4 and a number is 12 units to the right of -4 on a number line.

   $-4 + 12 = 8$

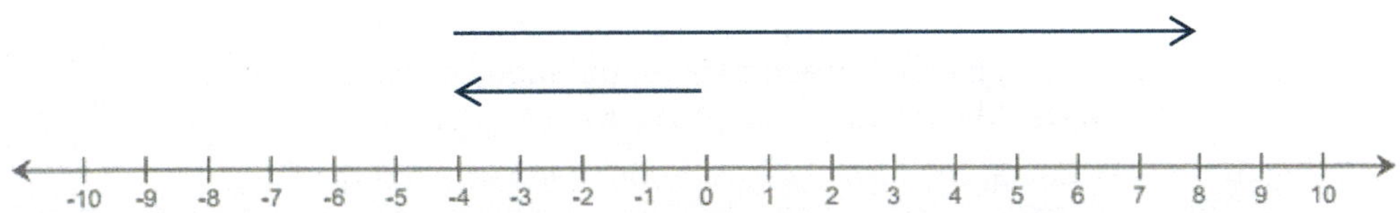

Lesson 4: Efficiently Adding Integers and Other Rational Numbers

Student Outcomes

- Students understand the rules for adding rational numbers:
  - Add rational numbers with the same sign by adding the absolute values and using the common sign.
  - Add rational numbers with opposite signs by subtracting the smaller absolute value from the larger absolute value and using the sign of the number with the larger absolute value.
- Students justify the rules using arrows and a number line or by using the Integer Game. They extend their findings to begin to include sums of rational numbers.

Classwork

Exercise 1 (5 minutes): Hands Up, Pair Up!

MP.3

Students review concepts from Lessons 1 through 3 by playing the Kagan Strategy Game, "Hands Up, Pair Up!"[1] During play, students should critique each other's questions when necessary. They should use accurate vocabulary learned so far in this module when explaining and defending their answers. The following are possible student questions:

- When playing the Integer Game, you have 3 cards in your hand with a sum of -15. Then, you draw a (-5) card. Using addition, how would you write an equation to represent your score?
- What is the absolute value of 15?
- What is the sum of $-4+(-10)$?
- In what direction does the arrow point on a number line for a negative number?
- What is the additive inverse of 5? What is the additive inverse of -9? What is the additive inverse of a number?

Scaffolding:

- Provide some premade index cards for learners who struggle forming a question with limited time.
- Ask students to refer to anchor posters for support during the game.

[1]Allow students 1–2 minutes to think of a question and record it on an index card. Write the answer to the question on the back. Ask the class to stand up, each person with one hand in the air. Students find partners and greet each other with a high five. Once a pair is formed, partners take turns asking each other their questions. After both partners have asked and answered each other's questions, they switch cards. Both partners again raise their hands to signify they are ready for a new partner and repeat the exercise. Allow enough time for each student to partner with 2–3 different people.

Example 1 (5 minutes): Rule for Adding Integers with Same Signs

The teacher leads the whole class to find the sum of $3 + 5$.

- In the Integer Game, I would combine 5 and 3 to give me 8.

Example 1: Rule for Adding Integers with Same Signs

a. **Represent the sum of $3 + 5$ using arrows on the number line.**

$3 + 5 = 8$

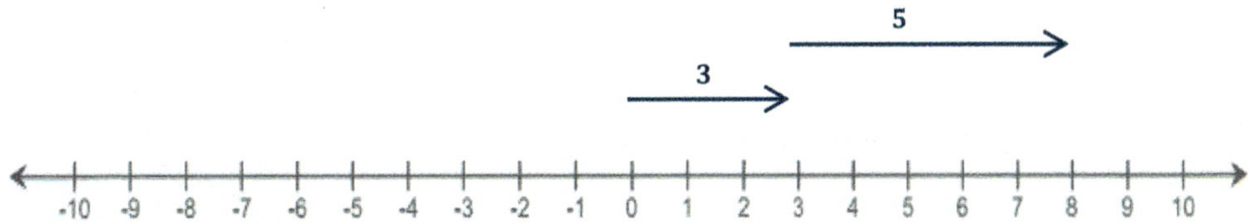

i. **How long is the arrow that represents 3?**

3 units

ii. **What direction does it point?**

Right/up

iii. **How long is the arrow that represents 5?**

5 units

iv. **What direction does it point?**

Right/up

v. **What is the sum?**

8

vi. **If you were to represent the sum using an arrow, how long would the arrow be, and what direction would it point?**

The arrow would be 8 units long and point to the right (or up on a vertical number line).

vii. **What is the relationship between the arrow representing the number on the number line and the absolute value of the number?**

The length of an arrow representing a number is equal to the absolute value of the number.

viii. **Do you think that adding two positive numbers will always give you a greater positive number? Why?**

Yes, because the absolute values are positive, so the sum will be a greater positive. On a number line, adding them would move you farther away from 0 (to the right or above) on a number line.

Scaffolding:

- Provide premade number lines for use throughout the lesson.
- Introduce questions one at a time using projection technology to support nonauditory learners.
- Use polling software throughout the lesson to gauge the entire class's understanding.

From part (a), use the same questions to elicit feedback.

- In the Integer Game, I would combine -3 and -5 to give me -8.

b. Represent the sum of $-3 + (-5)$ using arrows that represent -3 and -5 on the number line.

$-3 + (-5) = -8$

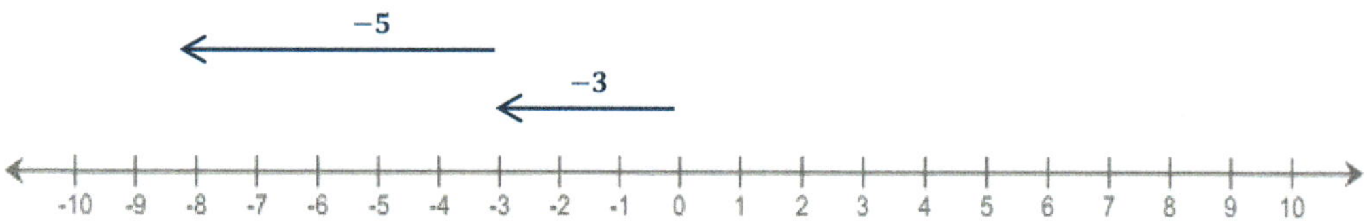

i. How long is the arrow that represents -3?

3 units

ii. What direction does it point?

Left/down

iii. How long is the arrow that represents -5?

5 units

iv. What direction does it point?

Left/down

v. What is the sum?

-8

vi. If you were to represent the sum using an arrow, how long would the arrow be, and what direction would it point?

The arrow would be 8 units long and point to the left (or down on a vertical number line).

vii. Do you think that adding two negative numbers will always give you a smaller negative number? Why?

Yes, because the absolute values of negative numbers are positive, so the sum will be a greater positive. However, the opposite of a greater positive is a smaller negative. On a number line, adding two negative numbers would move you farther away from 0 (to the left or below) on a number line.

c. What do both examples have in common?

The length of the arrow representing the sum of two numbers with the same sign is the same as the sum of the absolute values of both numbers.

The teacher writes the rule for adding integers with the same sign.

> ***RULE:*** ***Add rational numbers with the same sign by adding the absolute values and using the common sign.***

Exercise 2 (5 minutes)

Students work in pairs while solving practice problems.

> *Scaffolding:*
> - Create anchor posters when introducing integer addition rules (i.e., Adding Same Sign and Adding Opposite Signs).
> - Use a gallery wall to post examples and generate student discussion.

Exercise 2

a. **Decide whether the sum will be positive or negative without actually calculating the sum.**

i. $-4 + (-2)$ ***Negative***

ii. $5 + 9$ ***Positive***

iii. $-6 + (-3)$ ***Negative***

iv. $-1 + (-11)$ ***Negative***

v. $3 + 5 + 7$ ***Positive***

vi. $-20 + (-15)$ ***Negative***

b. **Find the sum.**

i. $15 + 7$

22

ii. $-4 + (-16)$

-20

iii. $-18 + (-64)$

-82

iv. $-205 + (-123)$

-328

Example 2 (7 minutes): Rule for Adding Opposite Signs

The teacher leads the whole class to find the sum of $5 + (-3)$.

- In the Integer Game, I would combine 5 and -3 to give me 2.

Example 2: Rule for Adding Opposite Signs

a. **Represent $5 + (-3)$ using arrows on the number line.**

$5 + (-3) = 2$

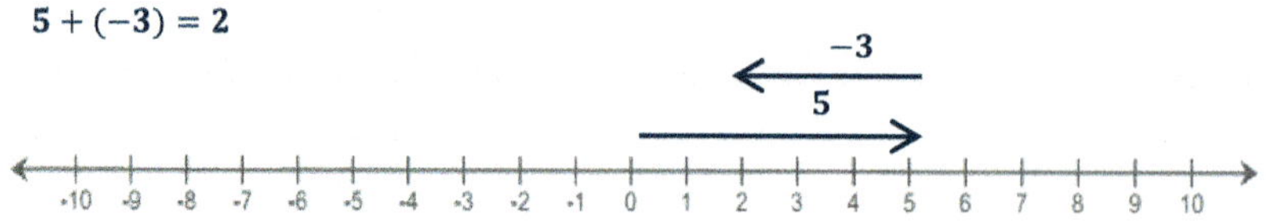

i. **How long is the arrow that represents 5?**

5 *units*

ii. **What direction does it point?**

Right/up

iii. **How long is the arrow that represents -3?**

3 *units*

iv. **What direction does it point?**

Left/down

v. **Which arrow is longer?**

The arrow that is five units long, or the first arrow.

vi. **What is the sum? If you were to represent the sum using an arrow, how long would the arrow be, and what direction would it point?**

$5 + (-3) = 2$

The arrow would be 2 *units long and point right/up.*

b. **Represent the $4 + (-7)$ using arrows on the number line.**

$4 + (-7) = -3$

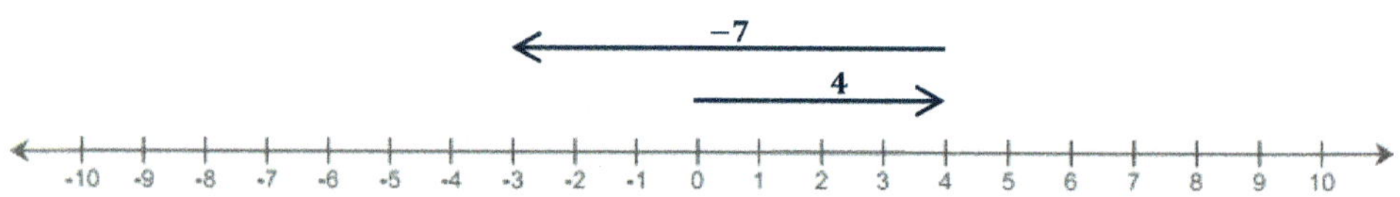

i. **In the two examples above, what is the relationship between the length of the arrow representing the sum and the lengths of the arrows representing the two addends?**

The length of the arrow representing the sum is equal to the difference of the absolute values of the lengths of both arrows representing the two addends.

ii. **What is the relationship between the direction of the arrow representing the sum and the direction of the arrows representing the two addends?**

The direction of the arrow representing the sum has the same direction as the arrow of the addend with the greater absolute value.

iii. **Write a rule that will give the length and direction of the arrow representing the sum of two values that have opposite signs.**

The length of the arrow of the sum is the difference of the absolute values of the two addends. The direction of the arrow of the sum is the same as the direction of the longer arrow.

The teacher writes the rule for adding integers with opposite signs.

***RULE:* Add rational numbers with opposite signs by subtracting the absolute values and using the sign of the integer with the greater absolute value.**

Exercise 3 (5 minutes)

Students work in pairs practicing addition with opposite signs. The teacher monitors student work and provides support when necessary.

Exercise 3

a. **Circle the integer with the greater absolute value. Decide whether the sum will be positive or negative without actually calculating the sum.**

i. $-1 + 2$ (circled: $+2$) *Positive*

ii. $5 + (-9)$ (circled: (-9)) *Negative*

iii. $-6 + 3$ (circled: -6) *Negative*

iv. $-11 + 1$ (circled: -11) *Negative*

b. **Find the sum.**

i. $-10 + 7$

-3

ii. $8 + (-16)$

-8

iii. $-12 + (65)$

53

iv. $105 + (-126)$

-21

G7-M2-B2-v1.3.0_05/04/15

Example 3 (5 minutes): Applying Integer Addition Rules to Rational Numbers

The teacher poses the example to the whole class. Students follow along in their student materials. The teacher poses additional questions to the class.

- Which addend has the greatest absolute value (length of the arrow)? What direction does this arrow point?
  - $|6| = 6$ *(The arrow length for 6 is 6 units long and to the right.)*
  - $\left|-2\frac{1}{4}\right| = 2\frac{1}{4}$ *(The arrow length for $-2\frac{1}{4}$ is $2\frac{1}{4}$ units long and to the left.)*
  - *The first addend has the greatest absolute value, and the arrow will point to the right.*
- What is the length of this arrow?
  - $|6| - \left|2\frac{1}{4}\right| = 3\frac{3}{4}$
- What is the final sign? What is the direction *of the resulting arrow?*
  - *Since 6 has the greater absolute value (arrow length), my answer will be positive, so* $6 + \left(-2\frac{1}{4}\right) = 3\frac{3}{4}$.

Example 3: Applying Integer Addition Rules to Rational Numbers

Find the sum of $6 + \left(-2\frac{1}{4}\right)$. The addition of rational numbers follows the same rules of addition for integers.

a. **Find the absolute values of the numbers.**

$|6| = 6$

$\left|-2\frac{1}{4}\right| = 2\frac{1}{4}$

b. **Subtract the absolute values.**

$6 - 2\frac{1}{4} = 6 - \frac{9}{4} = \frac{24}{4} - \frac{9}{4} = \frac{15}{4} = 3\frac{3}{4}$

c. **The answer will take the sign of the number that has the greater absolute value.**

Since 6 has the greater absolute value (arrow length), my answer will be positive $3\frac{3}{4}$.

Exercise 4 (5 minutes)

Students work independently while solving practice problems.

Exercise 4

Solve the following problems. Show your work.

a. **Find the sum of $-18 + 7$.**

$|-18| = 18$

$|7| = 7$

$18 - 7 = 11$

-11

b. If the temperature outside was 73 degrees at 5:00 p.m., but it fell 19 degrees by 10:00 p.m., what is the temperature at 10:00 p.m.? Write an equation and solve.

73 −19

$73 + (-19)$ $|73| = 73$

$73 - 19 = 54$ $|-19| = 19$

54

c. Write an addition sentence, and find the sum using the diagram below.

$-10 + 3\frac{1}{2} = -10 + \frac{7}{2} = -\frac{20}{2} + \frac{7}{2} = -\frac{13}{2} = -6\frac{1}{2}$ $-\left|\frac{20}{2}\right| = \frac{20}{2}$ $\left|\frac{7}{2}\right| = \frac{7}{2}$

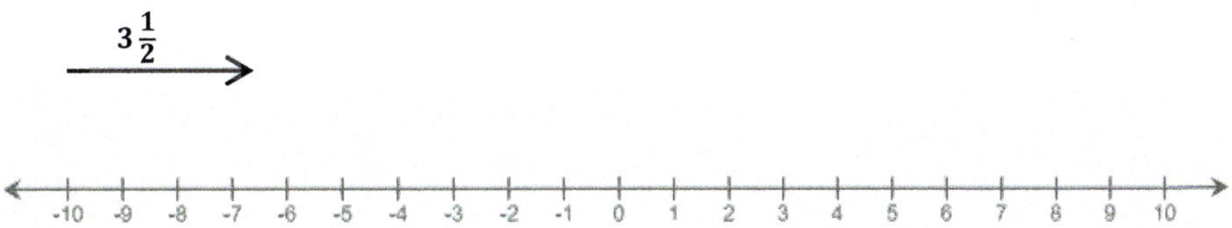

Closing (3 minutes)

The teacher calls on students at random to summarize the lesson.

- What are the rules of adding numbers with opposite signs?
  - *The rule for adding numbers with opposite signs is to subtract the absolute values, and then the answer will take the sign of the number with the largest absolute value.*
- What is the sum of $-3 + (-8)$?
  - *The sum of* $-3 + (-8)$ *is* -11.
- What do you think the rules would be for subtracting numbers with the same sign? (Do not spend too much time on this question. Allow students to verbally experiment with their responses.)
  - *Answers will vary.*

> *Scaffolding:*
>
> To help build confidence, allow students time to *turn and talk* with partners before posing questions.

> **Lesson Summary**
>
> - **Add integers with the same sign by adding the absolute values and using the common sign.**
> - **Steps to adding integers with opposite signs:**
>   1. **Find the absolute values of the integers.**
>   2. **Subtract the absolute values.**
>   3. **The answer will take the sign of the integer that has the greater absolute value.**
> - **To add rational numbers, follow the same rules used to add integers.**

Exit Ticket (5 minutes)

Name ______________________________ Date ______________

Lesson 4: Efficiently Adding Integers and Other Rational Numbers

Exit Ticket

1. Write an addition problem that has a sum of $-4\frac{3}{5}$ and

   a. The two addends have the same sign.

   b. The two addends have different signs.

2. In the Integer Game, what card would you need to draw to get a score of 0 if you have a -16, -35, and 18 in your hand?

Exit Ticket Sample Solutions

1. **Write an addition problem that has a sum of $-4\frac{4}{5}$ and**
   a. **The two addends have the same sign.**
      Answers will vary. $-1\frac{4}{5} + (-3) = -4\frac{4}{5}$.
   b. **The two addends have different signs.**
      Answers will vary. $1.8 + (-6.6) = -4.8$.

2. **In the Integer Game, what card would you need to draw to get a score of 0 if you have a -16, -35, and 18 in your hand?**
   $-16 + (-35) + 18 = -33$, *so I would need to draw a* 33 *because* 33 *is the additive inverse of* -33.
   $-33 + 33 = 0$.

Problem Set Sample Solutions

Students must understand the rules for addition of rational numbers with the same and opposite signs. The Problem Set presents multiple representations of these rules including diagrams, equations, and story problems. Students are expected to show their work or provide an explanation where necessary to justify their answers. Answers can be represented in fraction or decimal form.

1. **Find the sum. Show your work to justify your answer.**
   a. $4 + 17$
      $4 + 17 = 21$
   b. $-6 + (-12)$
      $-6 + (-12) = -18$
   c. $2.2 + (-3.7)$
      $2.2 + (-3.7) = -1.5$
   d. $-3 + (-5) + 8$
      $-3 + (-5) + 8 = -8 + 8 = 0$
   e. $\frac{1}{3} + \left(-2\frac{1}{4}\right)$
      $\frac{1}{3} + \left(-2\frac{1}{4}\right) = \frac{1}{3} + \left(-\frac{9}{4}\right) = \frac{4}{12} + \left(-\frac{27}{12}\right) = -\frac{23}{12} = -1\frac{11}{12}$

2. Which of these story problems describes the sum $19 + (-12)$? Check all that apply. Show your work to justify your answer.

   ___*X*___ Jared's dad paid him $\$19$ for raking the leaves from the yard on Wednesday. Jared spent $\$12$ at the movie theater on Friday. How much money does Jared have left?

   ______ Jared owed his brother $\$19$ for raking the leaves while Jared was sick. Jared's dad gave him $\$12$ for doing his chores for the week. How much money does Jared have now?

   ___*X*___ Jared's grandmother gave him $\$19$ for his birthday. He bought $\$8$ worth of candy and spent another $\$4$ on a new comic book. How much money does Jared have left over?

3. Use the diagram below to complete each part.

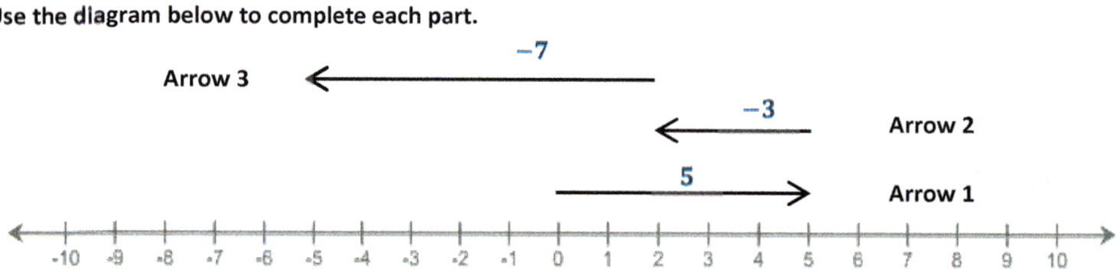

   a. Label each arrow with the number the arrow represents.

   b. How long is each arrow? What direction does each arrow point?

Arrow	Length	Direction
1	5	*right*
2	3	*left*
3	7	*left*

   c. Write an equation that represents the sum of the numbers. Find the sum.

   $5 + (-3) + (-7) = -5$

4. Jennifer and Katie were playing the Integer Game in class. Their hands are represented below.

Jennifer's Hand: 5, -8

Katie's Hand: -9, 7

   a. What is the value of each of their hands? Show your work to support your answer.

   Jennifer's hand has a value of -3 *because* $5 + (-8) = -3$. *Katie's hand has a value of* -2 *because* $-9 + 7 = -2$.

   b. If Jennifer drew two more cards, is it possible for the value of her hand not to change? Explain why or why not.

   It is possible for her hand not to change. Jennifer could get any two cards that are the exact opposites such as (-2) *and* 2. *Numbers that are exact opposites are called additive inverses, and they sum to* 0. *Adding the number* 0 *to anything will not change the value.*

c. If Katie wanted to win the game by getting a score of 0, what card would she need? Explain.

Katie would need to draw a 2 *because the additive inverse of* -2 *is* 2. $-2+2=0$.

d. If Jennifer drew -1 and -2, what would be her new score? Show your work to support your answer.

Jennifer's new score would be -6 *because* $-3+(-1)+(-2)=-6$.

Lesson 5: Understanding Subtraction of Integers and Other Rational Numbers

Student Outcomes

- Students justify the rule for subtraction: Subtracting a number is the same as adding its opposite.
- Students relate the rule for subtraction to the Integer Game: Removing (subtracting) a positive card changes the score in the same way as adding a corresponding negative card. Removing (subtracting) a negative card makes the same change as adding the corresponding positive card.

  Students justify the rule for subtraction for all rational numbers from the inverse relationship between addition and subtraction; that is, subtracting a number and adding it back gets you back to where you started: $(m - n) + n = m$.

Classwork

Example 1 (7 minutes): Exploring Subtraction with the Integer Game

Students play the Integer Game[1] in groups of 3–4, recording what happens in their student materials as they select and discard cards from their hand. Students use their previous knowledge of adding integers of same and opposite signs to help look for patterns when subtracting integers. In this example, students start with the cards 10, -2, and 4. The X indicates the cards that are removed from the hand.

Example 1: Exploring Subtraction with the Integer Game

Play the Integer Game in your group. Start Round 1 by selecting four cards. Follow the steps for each round of play.

1. **Write the value of your hand in the Total column.**
2. **Then, record what card values you select in the Action 1 column and discard from your hand in the Action 2 column.**
3. **After each action, calculate your new total, and record it under the appropriate Results column.**
4. **Based on the results, describe what happens to the value of your hand under the appropriate Descriptions column. For example, "Score increased by 3."**

[1]Refer to the Integer Game outline for player rules.

Sample student work.

Round	Total	Action 1	Result 1	Description	Action 2	Result 2	Description
1	12	*Select* 3	15	*Score (total) increased by* 3	*Discard* 10	5	*Score (total) decreased by* 10
2	5	*Select* −1	4	*Score (total) decreased by* 1	*Discard* 4	0	*Score (total) decreased by* 4
3	0	*Select* −7	−7	*Score (total) decreased by* 7	*Discard* −2	−5	*Score (total) increased by* 2
4	−5	*Select* 1	−4	*Score (total) increased by* 1	*Discard* −7	3	*Score (total) increased by* 7
5	3	*Select* 6	9	*Score (total) increased by* 9	*Discard* 1	8	*Score (total) decreased by* 1

Discussion (5 minutes): Making Connections to Integer Subtraction

The teacher leads the class in a discussion. The objective of the discussion is to allow students the opportunity to discuss any patterns they noticed while playing the game; in particular, what happens to the value of the hand when cards with negative values are selected or discarded. The teacher poses questions to individual groups to elicit student feedback.

Discussion: Making Connections to Integer Subtraction

1. **How did selecting positive value cards change the value of your hand?**

   It increased my score by the value of the card.

2. **How did selecting negative value cards change the value of your hand?**

   It decreased my score by the absolute value of the card.

3. **How did discarding positive value cards change the value of your hand?**

   It decreased my score by the value of the card.

4. **How did discarding negative value cards change the value of your hand?**

   It increased my score by the absolute value of the card.

5. **What operation reflects selecting a card?**

   Addition

6. **What operation reflects discarding or removing a card?**

   Subtraction

7. **Based on the game, can you make a prediction about what happens to the result when**
   a. **Subtracting a positive integer?**

      The result of the hand will decrease by the value of the integer.

   b. **Subtracting a negative integer?**

      The result of the hand will increase by the absolute value of the negative integer.

At the end of the lesson, the class reviews its predictions.

Scaffolding:

- Display questions and give students time to discuss in their groups prior to whole-class discussion.
- Allow students to use white boards, number lines, or tables to formulate and justify their opinions to the group.
- Record selected student responses and examples on chart paper to help identify patterns.

Example 2 (5 minutes): Subtracting a Positive Number

The teacher leads the whole class by modeling an Integer Game example to find the sum of $4 + 2$.

> *Scaffolding:*
>
> Allow students to use their Integer Cards throughout this example.

Example 2: Subtracting a Positive Number

Follow along with your teacher to complete the diagrams below.

If I had these two cards, the sum would be 6.

4 | 2

$4 + 2 = \boxed{6}$

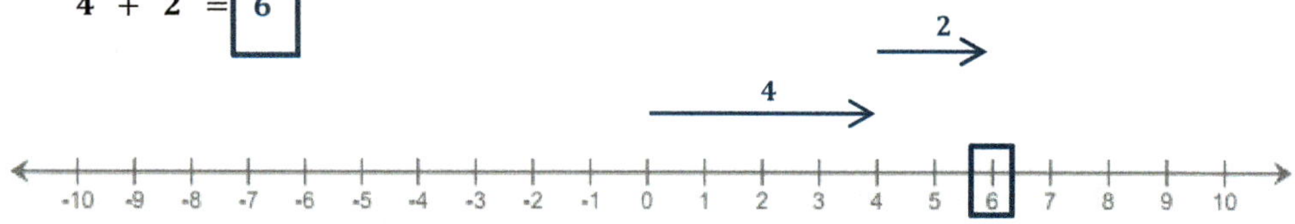

Show that discarding (subtracting) a positive card, which is the same as subtracting a positive number, decreases the value of the hand.

If I discarded or removed the 2, my score would decrease by 2 because I would still have a 4 left in my hand.

$4 + 2 - 2 = 4$. *Taking away, or subtracting, 2 causes my score to decrease by 2.*

4 | ~~2~~

Subtract (remove) the two.

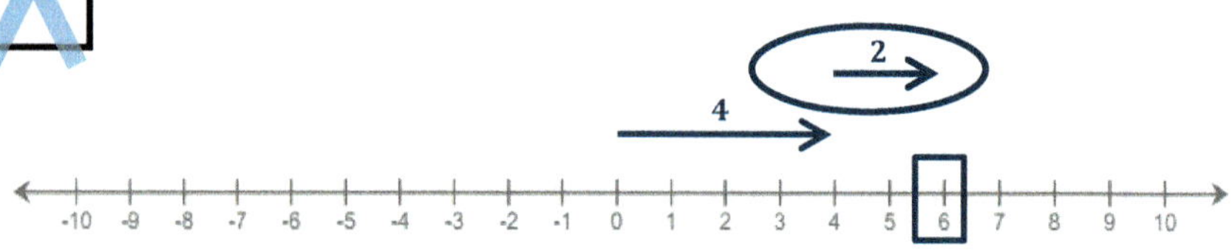

$4 + 2 - 2 = \boxed{4}$

OR

4 | 2 | −2

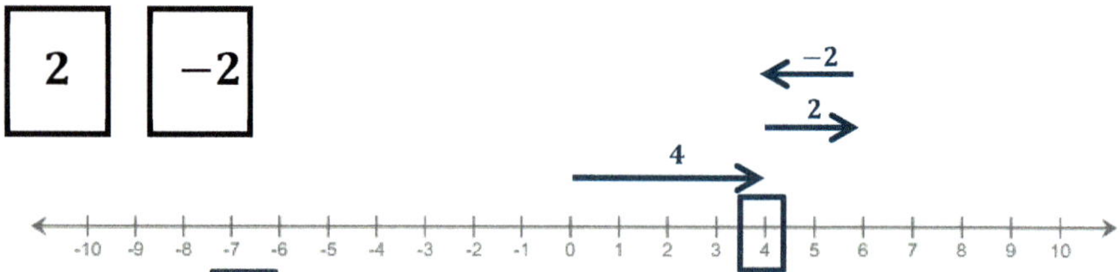

$4 + 2 + (-2) = \boxed{4}$

Removing (*subtracting*) a positive card changes the score in the same way as *adding* a card whose value is the *additive inverse* (or opposite). In this case, adding the corresponding *negative such that* $4 - 2 = 4 + (-2)$.

Subtracting a positive q-value is represented on the number line as moving to the left on a number line.

Example 3 (7 minutes): Subtracting a Negative Number

The teacher leads the whole class by modeling an Integer Game example to find the sum of $4 + (-2)$.

Example 3: Subtracting a Negative Number

Follow along with your teacher to complete the diagrams below.

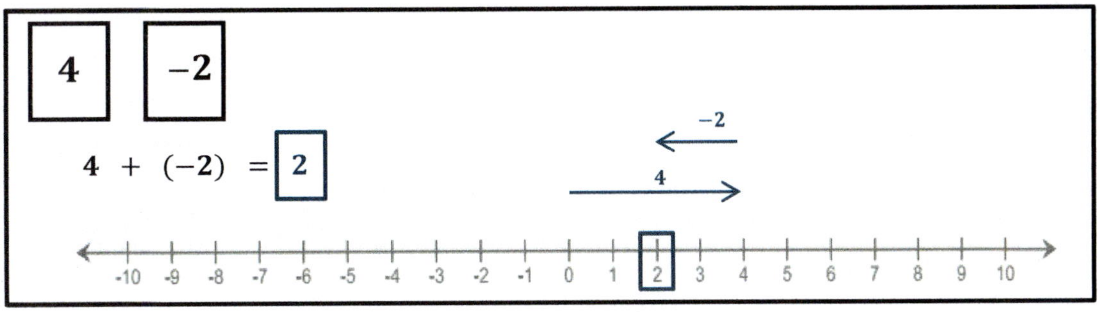

How does removing a negative card change the score, or value, of the hand?

If I discarded, or removed, the -2*, my score would increase by* 2 *because I would still have a* 4 *left in my hand.*

$4 + (-2) - (-2) = 4$. *Taking away, or subtracting,* -2 *causes my score to increase by* 2.

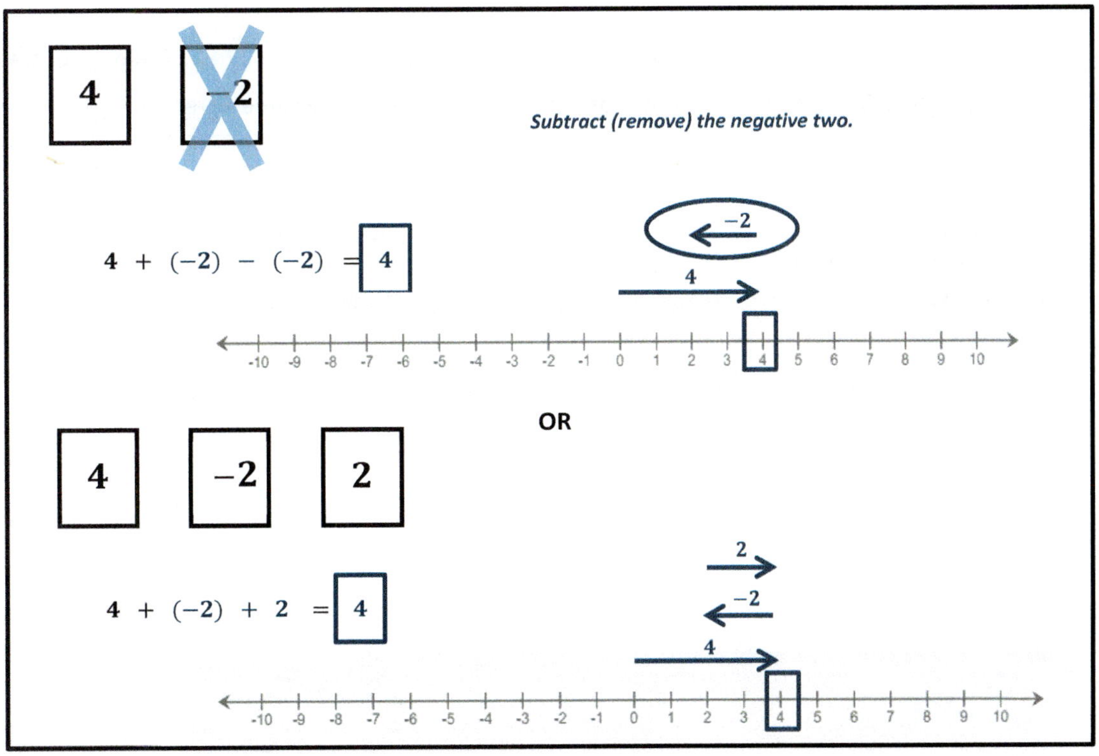

Removing (*subtracting*) a negative card changes the score in the same way as *adding* a card whose value is the *additive inverse* (or opposite). In this case, adding the corresponding *positive such that* $4 - (-2) = 4 + 2$.

Subtracting a negative q-value is represented on the number line as moving to the right on a number line because it is the opposite of subtracting a positive q-value (move to the left).

THE RULE OF SUBTRACTION: ***Subtracting a number is the same as adding its additive inverse (or opposite).***

Exercises 1–3 (8 minutes): Subtracting Positive and Negative Integers

Students work independently to find the differences below. Students may use the number line as additional support. Before students work on the exercises, model the examples below to help students make the connection between subtraction and addition of the additive inverse.

- To solve the problem $8 - 12$

$8 + 12$ (+ circled)	Step 1:	Change the subtraction sign to addition. (Rule of subtraction)
$8 + (-12)$ (−12 circled)	Step 2:	Change the positive 12 to a negative 12. (Rule of subtraction)
$\lvert 8\rvert = 8;\ \lvert -12\rvert = 12$		
	Steps 3–5:	Follow the steps for adding numbers with opposite signs.
$12 - 8 = 4$		Subtract the absolute values.
-4		Take the sign of the number with the greater absolute value.

Scaffolding:

Have students circle the integer with the greater absolute value to determine the final sign of the integer.

- Likewise, to solve the problem $4 - (-2)$

$4 + 2$	Step 1:	Change the subtraction sign to addition and the -2 to 2. (Rule of subtraction)
$4 + 2 = 6$	Step 2:	Follow the steps for adding numbers with the same signs.

Exercises 1–3: Subtracting Positive and Negative Integers

1. **Using the rule of subtraction, rewrite the following subtraction sentences as addition sentences and solve. Use the number line below if needed.**

   a. $8 - 2$

   $8 + (-2) = 6$

   b. $4 - 9$

   $4 + (-9) = -5$

   c. $-3 - 7$

   $-3 + (-7) = -10$

   d. $-9 - (-2)$

   $-9 + 2 = -7$

2. Find the differences.

a. $-2-(-5)$

$-2+5=3$

b. $11-(-8)$

$11+8=19$

c. $-10-(-4)$

$-10+4=-6$

3. Write two equivalent expressions that represent the situation. What is the difference in their elevations?

An airplane flies at an altitude of $25,000$ feet. A submarine dives to a depth of 600 feet below sea level.

$25,000-(-600)$ *and* $25,000+600$

The difference in their elevation is $25,600$ *feet.*

Closing (6 minutes)

Summarize the rules for subtracting rational numbers by posing the following questions to the class.

- Review your predictions made earlier in class. Were you correct? If not, how were your predictions different from the correct responses?
  - *Answers will vary.*
- When playing the Integer Game, give two ways you can increase the value of your hand.
  - *To increase the value of your hand during the Integer Game, you can either pick up a positive card or discard a negative card.*
- Give two ways you can decrease the value of your hand.
  - *To decrease the value of your hand during the Integer Game, you can either pick up a negative card or discard a positive card.*

Lesson Summary

- **THE RULE OF SUBTRACTION: Subtracting a number is the same as adding its opposite.**
- **Removing (subtracting) a positive card changes the score in the same way as adding a corresponding negative card.**
- **Removing (subtracting) a negative card makes the same change as adding the corresponding positive card.**
- **For all rational numbers, subtracting a number and adding it back gets you back to where you started:** $(m-n)+n=m$.

Exit Ticket (7 minutes)

Name ______________________________ Date ______________

Lesson 5: Understanding Subtraction of Integers and Other Rational Numbers

Exit Ticket

1. If a player had the following cards, what is the value of his hand?

1	-7	4

   a. Identify two different ways the player could get to a score of 5 by adding or removing only one card. Explain.

   b. Write two equations for part (a), one for each of the methods you came up with for arriving at a score of 5.

2. Using the rule of subtraction, rewrite the following subtraction expressions as addition expressions, and find the sums.

   a. $5 - 9$

   b. $-14 - (-2)$

Exit Ticket Sample Solutions

1. If a player had the following cards, what is the value of his hand?

   The current value of the hand is -2. $1+(-7)+4=-2$.

1	−7	4

   a. Identify two different ways the player could get to a score of 5 by adding or removing only one card. Explain.

   He could remove the -7 *or add* 7. *If he removes the* -7*, the value of the hand will be* 5*, which is* 7 *larger than* -2. *He could also get a sum of* 5 *by adding* 7 *to the hand. Therefore, removing the* -7 *gives him the same result as adding* 7.

   b. Write two equations for part (a), one for each of the methods you came up with for arriving at a score of 5.

   $-2-(-7)=5$ *and* $-2+7=5$

2. Using the rule of subtraction, rewrite the following subtraction expressions as addition expressions, and find the sums.

   a. $5-9$

   $5+(-9)=-4$

   b. $-14-(-2)$

   $-14+2=-12$

Problem Set Sample Solutions

The Problem Set provides students with skill practice and application of the rules for integer subtraction. Students solve problems with and without a number line.

1. On a number line, find the difference of each number and 4. Complete the table to support your answers. The first example is provided.

Number	Subtraction Expression	Addition Expression	Answer
10	$10-4$	$10+(-4)$	6
2	$2-4$	$2+(-4)$	-2
-4	$-4-4$	$-4+(-4)$	-8
-6	$-6-4$	$-6+(-4)$	-10
1	$1-4$	$1+(-4)$	-3

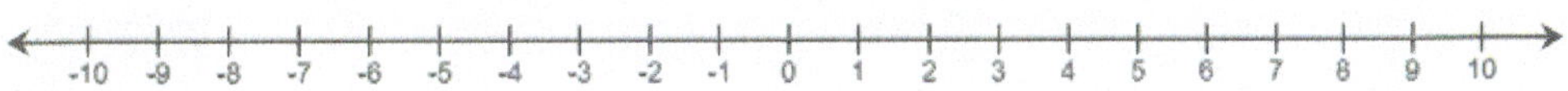

2. You and your partner were playing the Integer Game in class. Here are the cards in both hands.

Your hand

-8	6	1	-2

Your partner's hand

9	-5	2	-7

a. Find the value of each hand. Who would win based on the current scores? (The score closest to 0 wins.)

My hand: $-8 + 6 + 1 + (-2) = -3$

Partner's hand: $9 + (-5) + 2 + (-7) = -1$

My partner would win because -1 *is closer to* 0*. It is* 1 *unit to the left of* 0*.*

b. Find the value of each hand if you discarded the -2 and selected a 5, and your partner discarded the -5 and selected a 5. Show your work to support your answer.

My hand: Discard the -2, $-3 - (-2) = -1$*; Select a* 5*:* $-1 + 5 = 4$.

Partner's hand: Discard the -5, $-1 - (-5) = 4$*; Select a* 5*:* $4 + 5 = 9$.

c. Use your score values from part (b) to determine who would win the game now.

I would win now because 4 *is closer to zero.*

3. Write the following expressions as a single integer.

a. $-2 + 16$

14

b. $-2 - (-16)$

14

c. $18 - 26$

-8

d. $-14 - 23$

-37

e. $30 - (-45)$

75

4. **Explain what is meant by the following, and illustrate with an example:**

"For any real numbers, p and q, $p - q = p + (-q)$."

Subtracting a number is the same as adding its additive inverse. Examples will vary. A sample response is shown below.

$p = 4$, $q = 6$, $4 - 6$ *is the same as* $4 + (-6)$ *because* -6 *is the opposite of* 6.

$4 - 6 = -2$

$4 + (-6) = -2$

So, $4 - 6 = 4 + (-6)$ *because they both equal* -2.

5. **Choose an integer between -1 and -5 on the number line, and label it point P. Locate and label the following points on the number line. Show your work.**

Answers will vary. A sample response is shown below given that the student chose -3 *for* P.

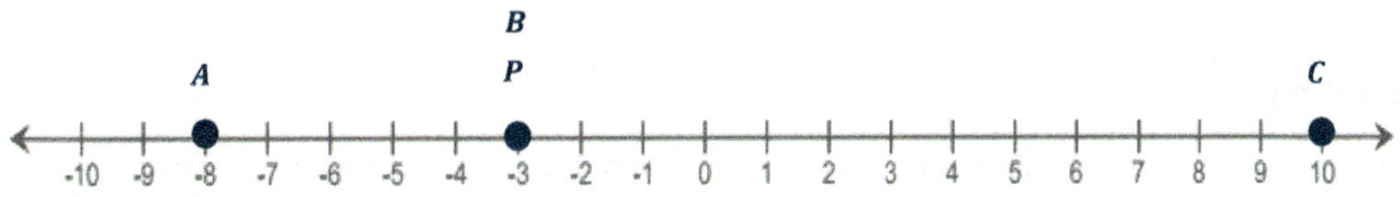

a. **Point A: $P - 5$**

Point A: $-3 - 5 = -8$

b. **Point B: $(P - 4) + 4$**

Point B: $(-3 - 4) + 4 = -3$ *(same as P)*

c. **Point C: $-P - (-7)$**

Point C: $-(-3) - (-7) = 3 + 7 = 10$

Challenge Problem:

6. **Write two equivalent expressions that represent the situation. What is the difference in their elevations?**

An airplane flies at an altitude of $26,000$ feet. A submarine dives to a depth of 700 feet below sea level.

Two equivalent expressions are $26,000 - (-700)$ *and* $26,000 + 700$. *The difference in their elevations is* $26,700$ *feet.*

Lesson 6: The Distance Between Two Rational Numbers

Student Outcomes

- Students justify the distance formula for rational numbers on a number line: If p and q are rational numbers on a number line, then the distance between p and q is $|p - q|$.
- Students know the definition of subtraction in terms of addition (i.e., $a - b = c$ means that $b + c = a$) and use the definition of subtraction to justify the distance formula.
- Students solve word problems involving changes in distance or temperature.

Classwork

Exercise 1 (4 minutes)

Students are in groups of 2; one person is Person A, and the other is Person B. Using a number line, each person independently counts the number of units that make up the distance between the two numbers listed in his assigned column. (The teacher may want to provide students with larger number lines.)

Exercise 1

Use the number line to answer each of the following.

Person A	Person B
What is the distance between -4 and 5? 9 	**What is the distance between 5 and -4?** 9 
What is the distance between -5 and -3? 2 	**What is the distance between -3 and -5?** 2 
What is the distance between 7 and -1? 8 	**What is the distance between -1 and 7?** 8 

After 3 minutes, partners share their answers and determine that their distances are the same because their endpoints are the same.

Note: A common mistake is that students count the first number as one. Another common mistake is that students describe the distance as negative.

Scaffolding:

Students may find it easier to see the distance if they use a highlighter on the number line and highlight the distance between the two numbers.

Discussion (5 minutes)

Follow-up Discussion

- What was the distance between -4 and 5?
  - 9
- What was the distance between 5 and -4?
  - 9
- Were you and your partner's answers the same for the second and third problems as well?
  - *Yes*
- Why did you both get the same answers for all three problems?
  - *Because the endpoints were the same, the distance between them is the same.*
- Take a minute to share with your partner a destination that you or your family usually travel to, for example, a relative's house or the location of an activity you attend each week. Assuming you take the same route to and from the location, what is the approximate distance in miles that you travel each way?
  - *Answers will vary, but students should recognize that the distance is the same to and from the location and should state the same number of miles whether traveling from home to the location or from the location back home.*

MP.5
- In life, at any given moment, will we always be able to use a number line to find the distance between two rational numbers? Is it the most efficient way to calculate distance between the two points?
  - *No*
- What represents the distance between a number and zero on the number line?
  - *Absolute value*
- If the distance between 5 and 0 can be calculated using $|5-0|$, or $|5|$, do you think we might be able to calculate the distance between -4 and 5 using absolute value? Take a minute to see if it works.
  - *Yes;* $|-4-5| = |-9| = 9$ *and* $|5-(-4)| = |9| = 9$*, which is the answer we found in Exercise 1.*
- Will this work for the other two distances we looked at in Exercise 1? Take a minute to test it out.
  - *Yes*

Exercise 2 (5 minutes)

Students now work independently using the formula to find the distance between each of the two given endpoints. They should verify their answer by using a number line model.

> *Scaffolding:*
> Consider having students determine the distance on the number line first and then use the formula to verify (rather than the other way around).

Exercise 2

Use the number line to answer each of the following questions.

a. What is the distance between 0 and -8?

$|0-(-8)| = |0+8| = |8| = 8$

b. What is the distance between -2 and $-1\frac{1}{2}$?

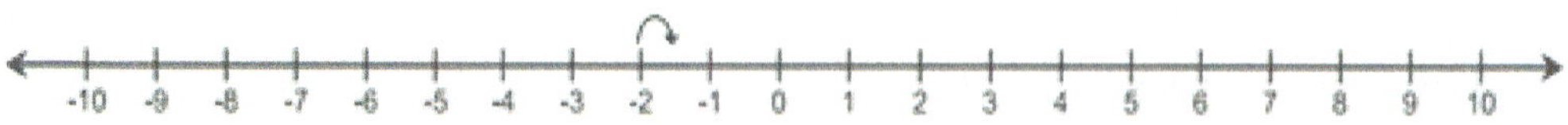

$$\left|-2-\left(-1\frac{1}{2}\right)\right| = \left|-2+1\frac{1}{2}\right| = \left|-\frac{1}{2}\right| = \frac{1}{2}$$

c. What is the distance between -6 and -10?

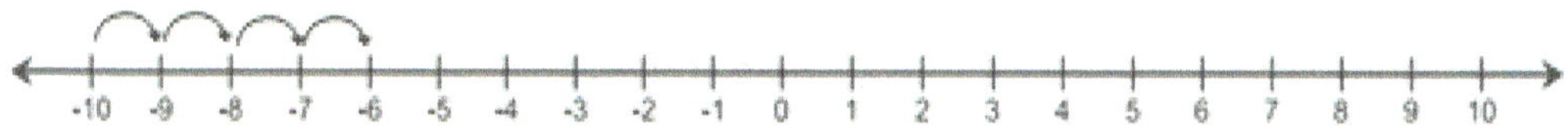

$$|-6-(-10)| = |-6+10| = |4| = 4$$

Scaffolding:

Before going over answers as a class, have students share their findings with their learning partner. They should justify their findings in words and be ready to share with the class.

Example 1 (3 minutes): Formula for the Distance Between Two Rational Numbers

If p and q are rational numbers on a number line, then the distance between p and q is $|p-q|$. It does not matter which endpoint we call p or which endpoint we call q. Distance is always positive.

Example 1: Formula for the Distance Between Two Rational Numbers

Find the distance between -3 and 2.

Step 1: Start on an endpoint.

Step 2: Count the number of units from the endpoint you started on to the other endpoint.

Using a formula,

$|-3-2| = |-3+-2| = |-5| = 5$ ***OR*** $|2-(-3)| = |2+3| = |5| = 5$

For two rational numbers p and q, the distance between p and q is $|p-q|$.

Example 2 (5 minutes): Change in Elevation vs. Distance

The teacher leads students in whole-group instruction; students record examples in their student materials.

Example 2: Change in Elevation vs. Distance

Distance is positive. Change in elevation or temperature may be positive or negative depending on whether it is increasing or decreasing (going up or down).

a. **A hiker starts hiking at the beginning of a trail at a point which is 200 feet below sea level. He hikes to a location on the trail that is 580 feet above sea level and stops for lunch.**

i. **What is the vertical distance between 200 feet below sea level and 580 feet above sea level?**

$|-200-580| = |-200+(-580)| = |-780| = 780$

The vertical distance is 780 feet.

ii. **How should we interpret 780 feet in the context of this problem?**

The hiker hiked 780 feet from a point below sea level to a point above sea level.

b. **After lunch, the hiker hiked back down the trail from the point of elevation, which is 580 feet above sea level, to the beginning of the trail, which is 200 feet below sea level.**

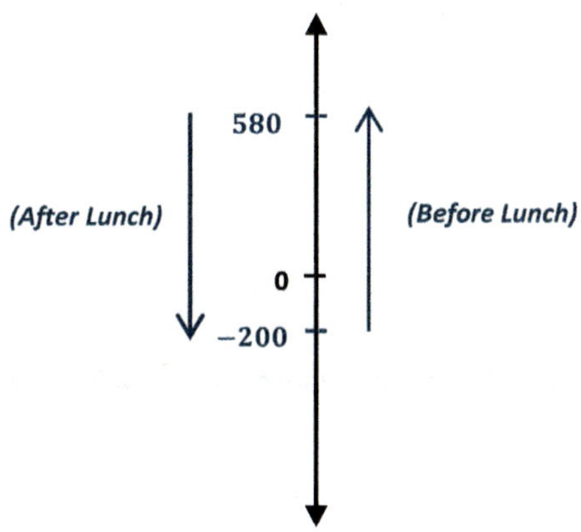

i. **What is the vertical distance between 580 feet above sea level and 200 feet below sea level?**

$|580-(-200)| = |580+200| = |780| = 780$

The vertical distance is 780 feet.

ii. **What is the change in elevation?**

780 feet

Exercise 3 (4 minutes)

Students work with a partner to come up with a solution to the following exercise. They use the distance formula to verify their answers but may first need to use a number line model to arrive at the two numbers for their solutions.

Note: A student may be tempted to use zero as one of the numbers; if that is the case, ask the student if zero is negative or positive.

Exercise 3

The distance between a negative number and a positive number is $12\frac{1}{2}$. What are the numbers?

Answers will vary; a possible solution is $-9\frac{1}{2}$ and 3. $\left|-9\frac{1}{2}-3\right| = \left|-9\frac{1}{2}+(-3)\right| = \left|-12\frac{1}{2}\right| = 12\frac{1}{2}$.

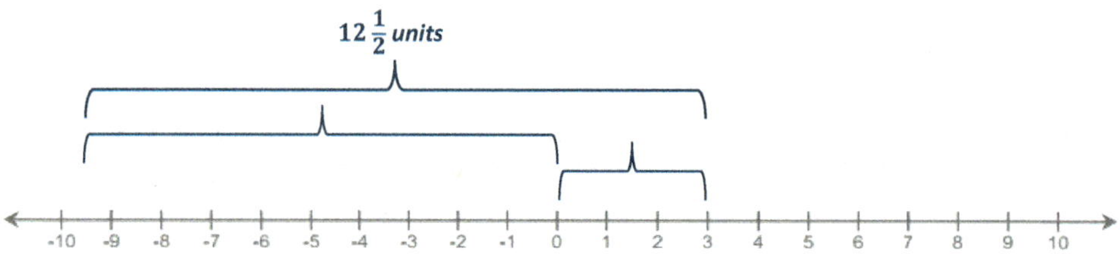

$\left|-9\frac{1}{2}\right| = 9\frac{1}{2}$ $\quad |3| = 3$

- Consider the different solutions students came up with, and notice that for each case if we take the absolute value of each of the endpoints, their sum is $12\frac{1}{2}$. Is that the case for Exercises 2 (a)–(c)? Why not?
  - *No, that is not always the case. If you have a positive number and a negative number as endpoints, they are separated by zero. And since absolute value is a number's distance from zero, taking the absolute value of each endpoint and adding them together will give you the same results as using the formula, $|p-q|$. However, if the signs of each endpoint are the same, you cannot calculate the distance between the two endpoints by finding the sum of their absolute values.*

Exercise 4 (10 minutes)

> *Scaffolding:*
>
> For parts (b)–(e), visual learners will benefit from using the number line to break down the distance into two sections, from zero to each given number.

Exercise 4

Use the distance formula to find each answer. Support your answer using a number line diagram.

a. **Find the distance between -7 and -4.**

$|-7-(-4)| = |-7+4| = |-3| = 3$

The distance between -7 and -4 is 3 units.

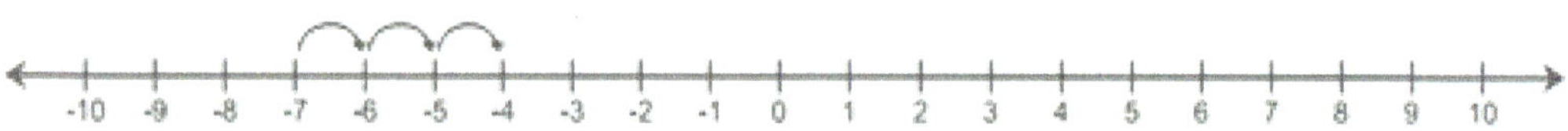

b. **Find the change in temperature if the temperature rises from -18°F to 15°F (use a vertical number line).**

$|15-(-18)| = |15+18| = |33| = 33$

The change in temperature is 33°F.

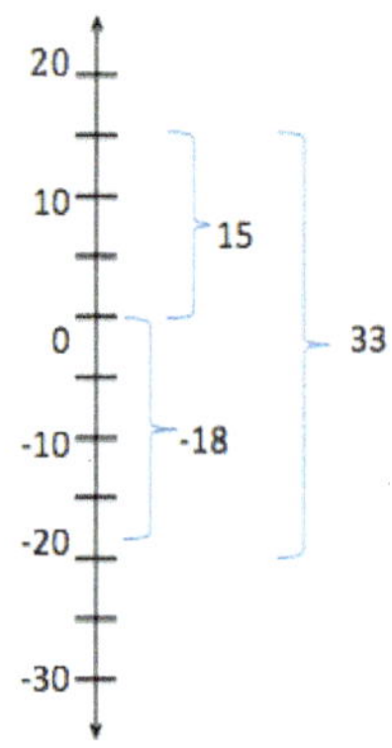

c. **Would your answer for part (b) be different if the temperature dropped from $15°\text{F}$ to $-18°\text{F}$? Explain.**

Yes. The distance between 15 and -18 on a number line is the same as the distance between -18 and 15 because the endpoints are the same, but the temperature would change by $-33°\text{F}$ because it is going from a higher value to a lower value.

d. **Beryl is the first person to finish a 5K race and is standing 15 feet beyond the finish line. Another runner, Jeremy, is currently trying to finish the race and has approximately 14 feet before he reaches the finish line. What is the minimum possible distance between Beryl and Jeremy?**

$|-14 - 15| = |-14 + (-15)| = |-29| = 29$

The minimum distance between Beryl and Jeremy is 29 feet.

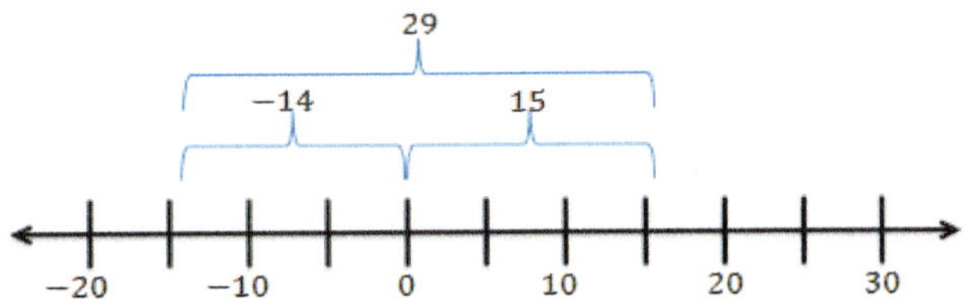

e. **What is the change in elevation from 140 feet above sea level to 40 feet below sea level? Explain.**

I used the distance formula: $|140 - (-40)| = 180$ and a vertical number line to find the distance between the two elevations. However, the elevation is going from a higher elevation to a lower one, so the change would be negative. Therefore, the change in elevation is -180 feet.

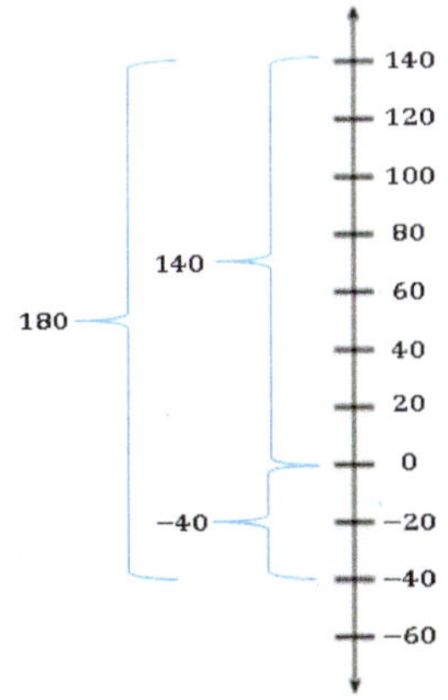

Closing (3 minutes)

- How can we use a number line to find the distance between two rational numbers?
  - *We can count the number of units in between the two numbers.*
- What does it mean to find the absolute value of a number?
  - *You are finding the distance between that number and zero.*
- Is it possible to use absolute value to find the distance between a number, p, and another number, q, that is not zero? If so, how?
  - *Yes, instead of using $|p - 0|$, you would use $|p - q|$.*
- Is distance always positive? Is change always positive?
  - *Distance is always positive, but change can be positive or negative.*

Lesson Summary

- **To find the distance between two rational numbers on a number line, you can count the number of units between the numbers.**
- **Using a formula, the distance between rational numbers, p and q, is $|p - q|$.**
- **Distance is always positive.**
- **Change may be positive or negative. For instance, there is a $-4°$ change when the temperature goes from $7°$ to $3°$.**

Exit Ticket (6 minutes)

Name ______________________________ Date ______________

Lesson 6: The Distance Between Two Rational Numbers

Exit Ticket

Two Grade 7 students, Monique and Matt, both solved the following math problem:

If the temperature drops from $7°\text{F}$ to $-17°\text{F}$, by how much did the temperature *decrease?*

The students came up with different answers. Monique said the answer is $24°\text{F}$, and Matt said the answer is $10°\text{F}$. Who is correct? Explain, and support your written response with the use of a formula and a vertical number line diagram.

Exit Ticket Sample Solutions

Two Grade 7 students, Monique and Matt, both solved the following math problem:

If the temperature drops from $7°\text{F}$ to $-17°\text{F}$, by how much did the temperature decrease?

The students came up with different answers. Monique said the answer is $24°\text{F}$, and Matt said the answer is $10°\text{F}$. Who is correct? Explain, and support your written response with the use of a formula and a vertical number line diagram.

Monique is correct. If you use the distance formula, you take the absolute value of the difference between 7 and -17 and that equals 24. Using a number line diagram, you can count the number of units between 7 and -17 to get 24.

$|7-(-17)| = |7+17| = |24| = 24$. *There was a $24°\text{F}$ drop in the temperature.*

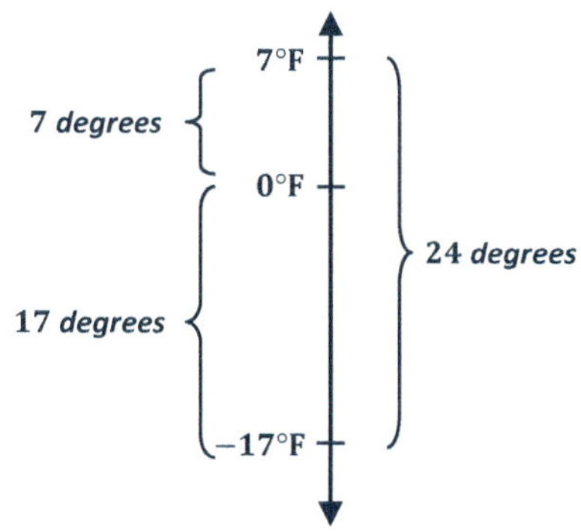

Problem Set Sample Solutions

1. $|-19-12| = |-19+(-12)| = |-31| = 31$
2. $|19-(-12)| = |19+12| = |31| = 31$
3. $|10-(-43)| = |10+43| = |53| = 53$
4. $|-10-43| = |-10+(-43)| = |-53| = 53$
5. $|-1-(-16)| = |-1+16| = |15| = 15$
6. $|1-16| = |1+(-16)| = |-15| = 15$
7. $|0-(-9)| = |0+9| = |9| = 9$
8. $|0-9| = |0+(-9)| = |-9| = 9$
9. $|-14.5-13| = |-14.5+(-13)| = |-27.5| = 27.5$
10. $|14.5-(-13)| = |14.5+13| = |27.5| = 27.5$

MP.7

11. **Describe any patterns you see in the answers to the problems in the left- and right-hand columns. Why do you think this pattern exists?**

    Each problem in the right-hand column has the same answer as the problem across from it in the left-hand column. That is because you are finding the distance between the opposite numbers as compared to the first column. The difference between the opposite numbers is opposite the difference between the original numbers. The absolute values of opposite numbers are the same.

Lesson 7: Addition and Subtraction of Rational Numbers

Student Outcomes

- Students recognize that the rules for adding and subtracting integers apply to rational numbers.
- Given a number line, students use arrows to model rational numbers where the length of the arrow is the absolute value of the rational number and the sign of the rational number is determined by the direction of the arrow with respect to the number line.
- Students locate the sum $p + q$ of two rational numbers on a number line by placing the tail of the arrow for q at p and locating $p + q$ at the head of the arrow. They create an arrow for the difference $p - q$ by first rewriting the difference as a sum, $p + (-q)$, and then locating the sum.

Classwork

Exercise 1 (5 minutes): Real-World Connection to Adding and Subtracting Rational Numbers

MP.4 Students answer the following question independently as the teacher circulates around the room providing guidance and feedback as needed. Students focus on how to represent the answer using both an equation and a number line diagram.

Exercise 1: Real-World Connection to Adding and Subtracting Rational Numbers

Suppose a seventh grader's birthday is today, and she is 12 years old. How old was she $3\frac{1}{2}$ years ago? Write an equation, and use a number line to model your answer.

$12 + \left(-3\frac{1}{2}\right) = 8\frac{1}{2}$ *or* $12 - 3\frac{1}{2} = 8\frac{1}{2}$

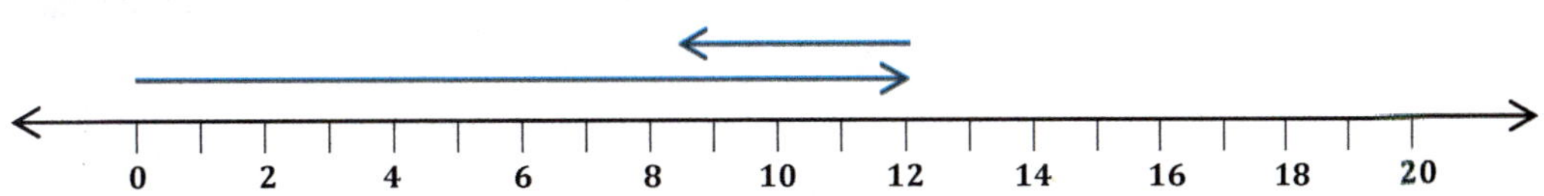

Example 1 (5 minutes): Representing Sums of Rational Numbers on a Number Line

MP.6 The teacher leads a whole-group instruction illustrating the sum of $12 + \left(-3\frac{1}{2}\right)$ on a number line. Elicit student responses to assist in creating the steps. Students record the steps and diagram.

Scaffolding:
Laminate an index card with the steps for Examples 1 and 2 and the number line diagram so that students can easily refer to it.

Example 1: Representing Sums of Rational Numbers on a Number Line

a. **Place the tail of the arrow on 12.**

b. **The length of the arrow is the absolute value of $-3\frac{1}{2}$, $\left|-3\frac{1}{2}\right| = 3\frac{1}{2}$.**

c. **The direction of the arrow is to the *left* since you are adding a negative number to 12.**

Draw the number line model in the space below.

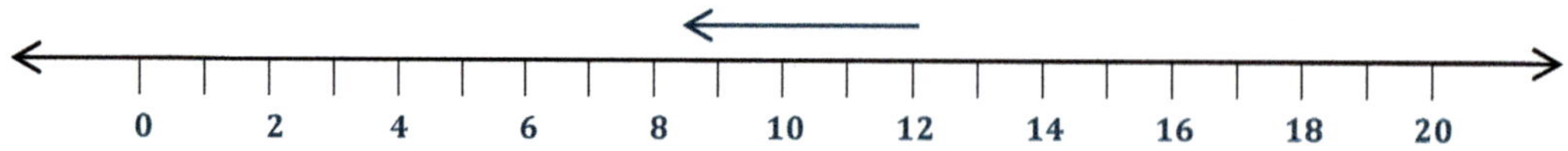

$12 + \left(-3\frac{1}{2}\right) = 8\frac{1}{2}$ *or* $12 - 3\frac{1}{2} = 8\frac{1}{2}$

Exercise 2 (3 minutes)

Exercise 2

Find the following sum using a number line diagram: $-2\frac{1}{2} + 5$.

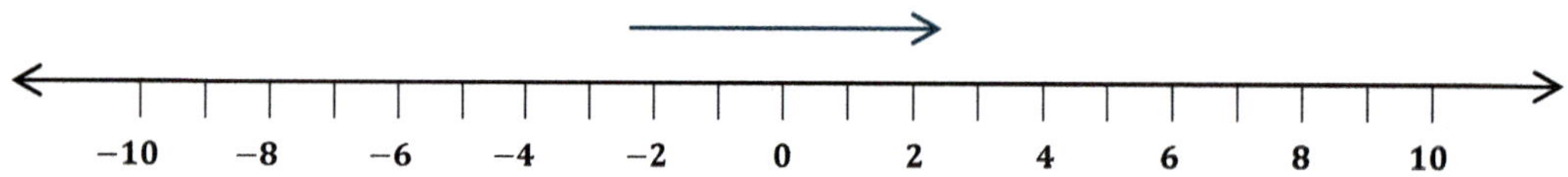

$\left(-2\frac{1}{2}\right) + 5 = 2\frac{1}{2}$

Example 2 (5 minutes): Representing Differences of Rational Numbers on a Number Line

The teacher leads a whole-group instruction illustrating how to find the difference of $1 - 2\frac{1}{4}$ on a number line. Elicit student responses to assist in creating the steps. Students record the steps and diagram.

> *Scaffolding:*
>
> Have students subdivide their number line intervals into $\frac{1}{2}$'s and $\frac{1}{4}$'s.

Example 2: Representing Differences of Rational Numbers on a Number Line

Find the following difference, and represent it on a number line: $1 - 2\frac{1}{4}$.

a. *Rewrite the difference $1 - 2\frac{1}{4}$ as a sum: $1 + \left(-2\frac{1}{4}\right)$.*

<u>Now follow the steps to represent the sum</u>:

b. *Place the tail of the arrow on 1.*

c. *The length of the arrow is the absolute value of $-2\frac{1}{4}$; $\left|-2\frac{1}{4}\right| = 2\frac{1}{4}$.*

d. *The direction of the arrow is to the left since you are adding a negative number to 1.*

Draw the number line model in the space below.

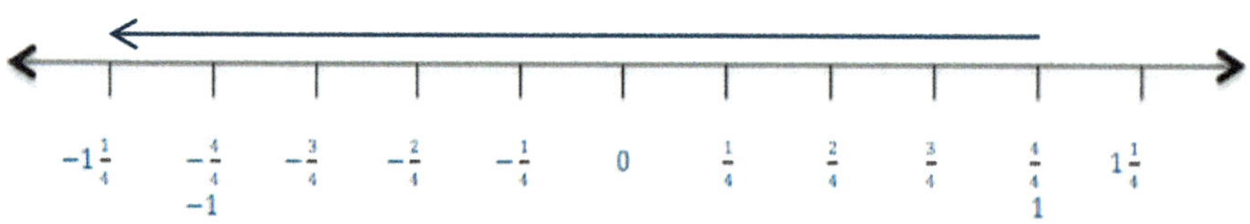

$1 + \left(-2\frac{1}{4}\right) = -1\frac{1}{4}$

Exercise 3 (3 minutes)

Exercise 3

Find the following difference, and represent it on a number line: $-5\frac{1}{2} - (-8)$.

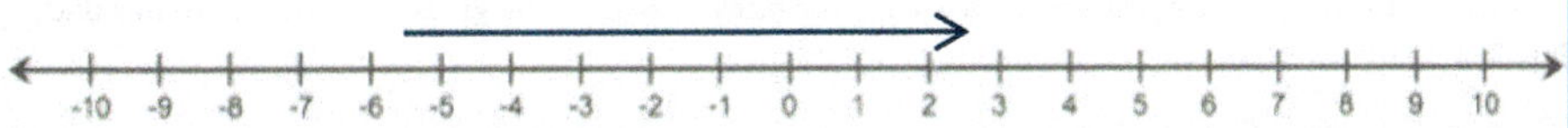

$\left(-5\frac{1}{2}\right) + 8 = 2\frac{1}{2}$

> *Scaffolding:*
> - Ask students to explain and justify what they drew to check for understanding.
> - Ask probing questions such as "Why does your arrow go to the right?"

Exercise 4 (10 minutes)

Next, students work independently in Exercise 4 to create a number line model to represent each sum or difference. After 5–7 minutes, students are selected to share their responses and work with the class.

Exercise 4

Find the following sums and differences using a number line model.

a. $-6 + 5\frac{1}{4}$

$-6 + 5\frac{1}{4} = -\frac{3}{4}$

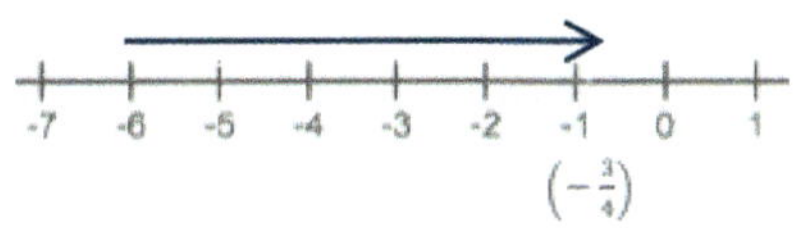

b. $7 - (-0.9)$

$7 + (0.9) = 7.9$

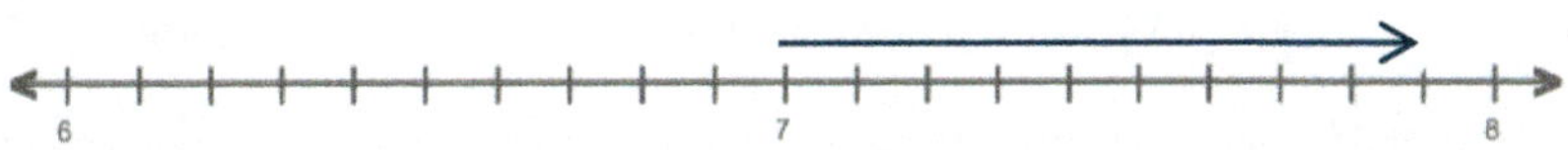

c. $2.5 + \left(-\frac{1}{2}\right)$

$2.5 + (-0.5) = 2$

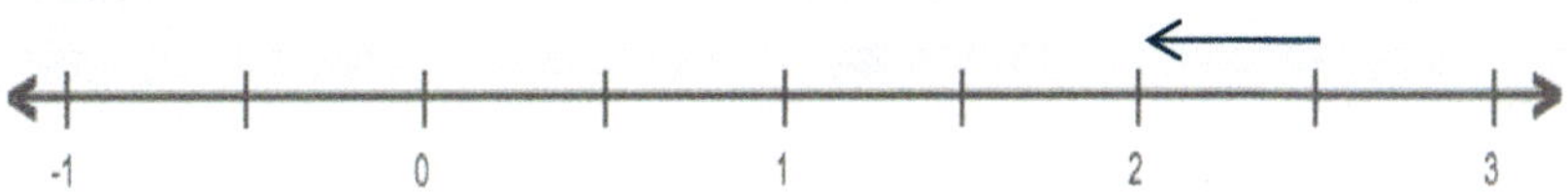

d. $-\frac{1}{4} + 4$

$-\frac{1}{4} + 4 = 3\frac{3}{4}$

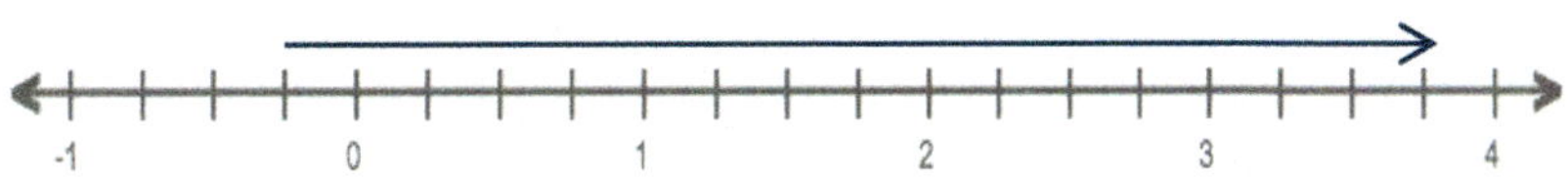

e. $\frac{1}{2} - (-3)$

$\frac{1}{2} + 3 = 3\frac{1}{2}$

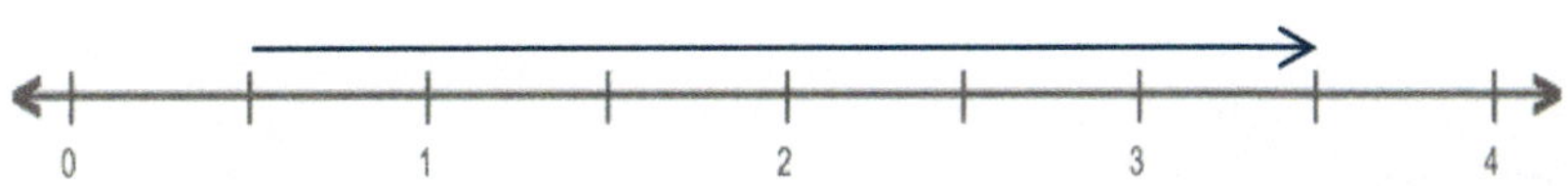

Exercise 5 (6 minutes)

Exercise 5

Create an equation and number line diagram to model each answer.

a. **Samantha owes her father $\$7$. She just got paid $\$5.50$ for babysitting. If she gives that money to her dad, how much will she still owe him?**

$-7 + 5.50 = -1.50$. *She still owes him* $\$1.50$.

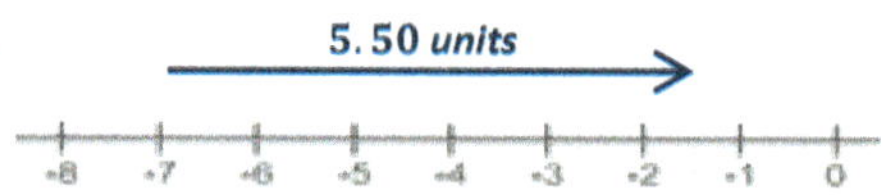

b. **At the start of a trip, a car's gas tank contains 12 gallons of gasoline. During the trip, the car consumes $10\frac{1}{8}$ gallons of gasoline. How much gasoline is left in the tank?**

$12 + \left(-10\frac{1}{8}\right) = 1\frac{7}{8}$ *or* $12 - 10\frac{1}{8} = 1\frac{7}{8}$. *There are* $1\frac{7}{8}$ *gallons of gas left in the tank.*

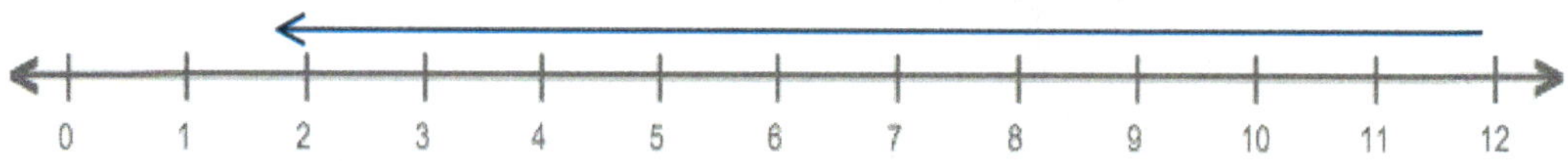

c. **A fish was swimming $3\frac{1}{2}$ feet below the water's surface at 7:00 a.m. Four hours later, the fish was at a depth that is $5\frac{1}{4}$ feet below where it was at 7:00 a.m. What rational number represents the position of the fish with respect to the water's surface at 11:00 a.m.?**

$-3\frac{1}{2} + \left(-5\frac{1}{4}\right) = -8\frac{3}{4}$. *The fish is* $8\frac{3}{4}$ *feet below the water's surface.*

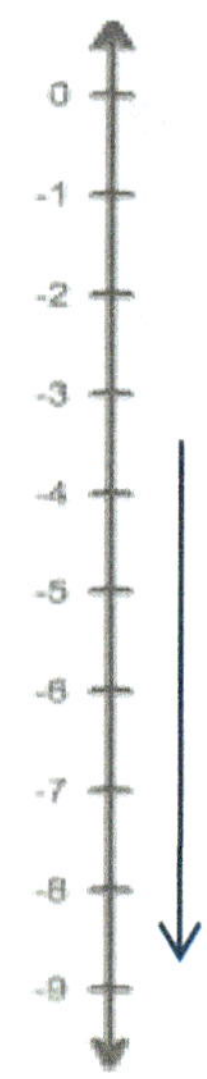

Follow-Up Discussion

For Exercise 5(a), discuss with students how the mathematical answer of -1.50 means Samantha owes her father $\$1.50$ and that we do not say she owes her father $-\$1.50$.

Closing (3 minutes)

- What challenges do you face when using the number line model to add non-integer rational numbers?
  - *Answers will vary.*
- When using a number line to model $8 - (-2.1)$, how many units do we move from 8 and in what direction? Where is the tail of the arrow, and where is the head? What does your arrow represent?
  - *First, we would change the expression to an addition expression,* $8 + 2.1$*. The tail of the arrow would start at* 8*, the first addend. The arrow would be* 2.1 *units long and pointing to the right, which would mean the arrow would end on* 10.1*. The arrow represents the second addend.*

> Lesson Summary
>
> **The rules for adding and subtracting integers apply to all rational numbers.**
>
> **The sum of two rational numbers (e.g., $-1 + 4.3$) can be found on the number line by placing the tail of an arrow at -1 and locating the head of the arrow 4.3 units to the right to arrive at the sum, which is 3.3.**
>
> **To model the difference of two rational numbers on a number line (e.g., $-5.7 - 3$), first rewrite the difference as a sum, $-5.7 + (-3)$, and then follow the steps for locating a sum. Place a single arrow with its tail at -5.7 and the head of the arrow 3 units to the left to arrive at -8.7.**

Exit Ticket (5 minutes)

Name ______________________________ Date______________

Lesson 7: Addition and Subtraction of Rational Numbers

Exit Ticket

At the beginning of the summer, the water level of a pond is 2 feet below its normal level. After an unusually dry summer, the water level of the pond dropped another $1\frac{1}{3}$ feet.

1. Use a number line diagram to model the pond's current water level in relation to its normal water level.

2. Write an equation to show how far above or below the normal water level the pond is at the end of the summer.

Exit Ticket Sample Solutions

At the beginning of the summer, the water level of a pond is 2 feet below its normal level. After an unusually dry summer, the water level of the pond dropped another $1\frac{1}{3}$ feet.

1. **Use a number line diagram to model the pond's current water level in relation to its normal water level.**

   Move $1\frac{1}{3}$ units to the left of -2. $-3\frac{1}{3}$

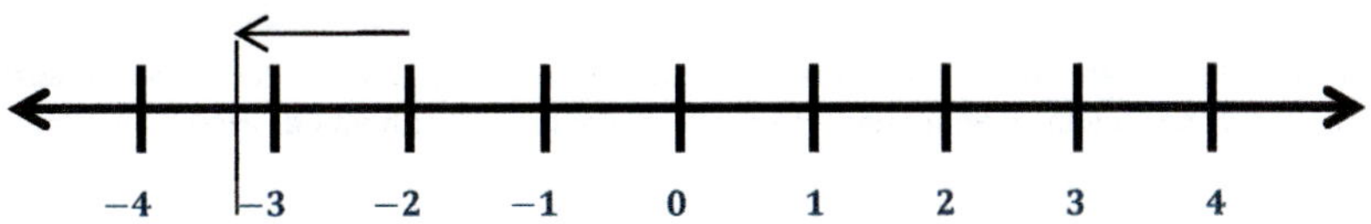

2. **Write an equation to show how far above or below the normal water level the pond is at the end of the summer.**

   $-2 - 1\frac{1}{3} = -3\frac{1}{3}$ *or* $-2 + \left(-1\frac{1}{3}\right) = -3\frac{1}{3}$

Problem Set Sample Solutions

Represent each of the following problems using both a number line diagram and an equation.

1. **A bird that was perched atop a $15\frac{1}{2}$-foot tree dives down six feet to a branch below. How far above the ground is the bird's new location?**

   $15\frac{1}{2} + (-6) = 9\frac{1}{2}$ *or* $15\frac{1}{2} - 6 = 9\frac{1}{2}$

   The bird is $9\frac{1}{2}$ feet above the ground.

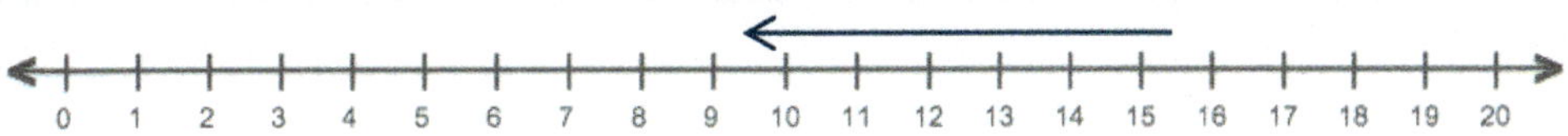

2. **Mariah owed her grandfather $\$2.25$ but was recently able to pay him back $\$1.50$. How much does Mariah currently owe her grandfather?**

   $-2.25 + 1.50 = -0.75$

   Mariah owes her grandfather 75 cents.

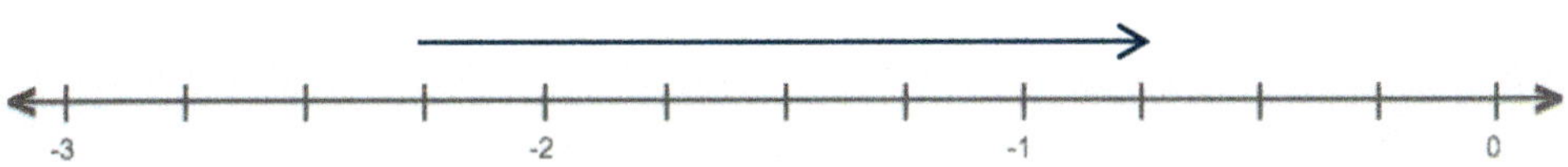

3. Jake is hiking a trail that leads to the top of a canyon. The trail is 4.2 miles long, and Jake plans to stop for lunch after he completes 1.6 miles. How far from the top of the canyon will Jake be when he stops for lunch?

$4.2 - 1.6 = 2.6$

Jake will be 2.6 miles from the top of the canyon.

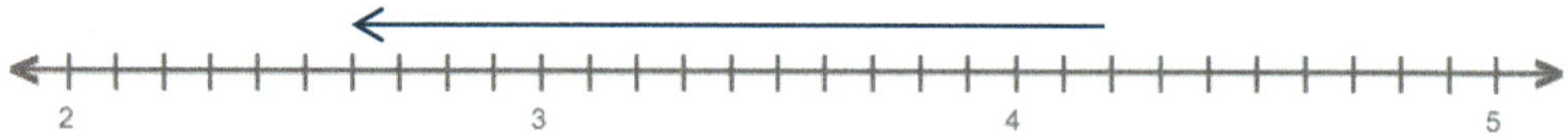

4. Sonji and her friend Rachel are competing in a running race. When Sonji is 0.4 miles from the finish line, she notices that her friend Rachel has fallen. If Sonji runs one-tenth of a mile back to help her friend, how far will she be from the finish line?

$-0.4 + (-0.1) = -0.5$ *or* $-0.4 - 0.1 = -0.5$

Sonji will be 0.5 miles from the finish line.

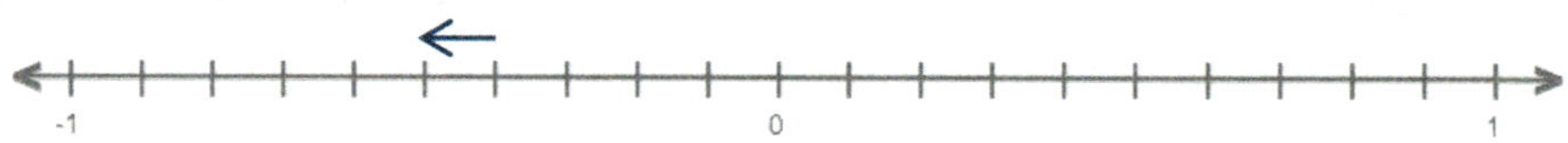

5. Mr. Henderson did not realize his checking account had a balance of $\$200$ when he used his debit card for a $\$317.25$ purchase. What is his checking account balance after the purchase?

$200 + (-317.25) = -117.25$ *or* $200 - 317.25 = -117.25$

Mr. Henderson's checking account balance will be $-\$117.25$.

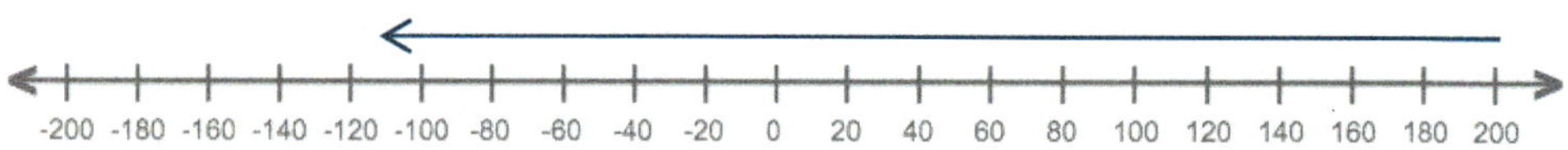

6. If the temperature is $-3°\text{F}$ at 10:00 p.m., and the temperature falls four degrees overnight, what is the resulting temperature?

$-3 - 4 = -3 + (-4) = -7$

The resulting temperature is $-7°\text{F}$.

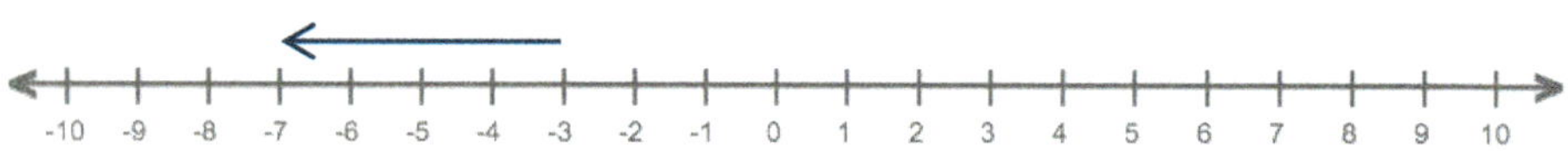

EUREKA MATH™

Lesson 8: Applying the Properties of Operations to Add and Subtract Rational Numbers

Student Outcomes

- Students use properties of operations to add and subtract rational numbers without the use of a calculator.
- Students recognize that any problem involving addition and subtraction of rational numbers can be written as a problem using addition and subtraction of positive numbers only.
- Students use the commutative and associative properties of addition to rewrite numerical expressions in different forms. They know that the opposite of a sum is the sum of the opposites (e.g., $-\big(3 + (-4)\big) = -3 + 4$).

Lesson Notes

This lesson is the first of a two-day lesson using the properties of operations to add and subtract rational numbers. The lesson begins with a focus on representing the opposite of a sum as the sum of its opposites so that students may more efficiently arrive at sums and differences of rational numbers. The focus includes a representation of negative mixed numbers so that students conceptualize a negative mixed number as a negative integer plus a negative fraction. Students often mistakenly add a negative mixed number to a positive whole number by adding the negative whole number part of the mixed number to the positive whole number but then erroneously representing the fractional part of the negative mixed number as a positive number.

The following is an example of the properties and how they are used in this lesson.

$-13\frac{5}{7} + 6 - \frac{2}{7}$

$= -13\frac{5}{7} + 6 + \left(-\frac{2}{7}\right)$ *Subtracting a number is the same as adding its inverse.*

$= -13 + \left(-\frac{5}{7}\right) + 6 + \left(-\frac{2}{7}\right)$ *The opposite of a sum is the sum of its opposite.*

$= -13 + \left(-\frac{5}{7}\right) + \left(-\frac{2}{7}\right) + 6$ *Commutative property of addition*

$= -13 + (-1) + 6$ *Associative property of addition*

$= -14 + 6$

$= -8$

Recall the Properties of Operations from Table 3 of the Common Core State Standards.

Table 3. The Properties of Operations. Here a, b, and c stand for arbitrary numbers.

Associative property of addition	$(a+b)+c=a+(b+c)$
Commutative property of addition	$a+b=b+a$
Additive identity property of 0	$a+0=0+a=a$
Existence of additive inverses	For every a there exists $-a$ so that $a+(-a)=(-a)+a=0$
Associative property of multiplication	$(a\cdot b)\cdot c=a\cdot(b\cdot c)$
Commutative property of multiplication	$a\cdot b=b\cdot a$
Multiplicative identity property of 1	$a\cdot 1=1\cdot a=a$
Existence of multiplicative inverses	For every $a\neq 0$ there exists $\frac{1}{a}$ so that $a\cdot\frac{1}{a}=\frac{1}{a}\cdot a=1$
Distributive property of multiplication over addition	$a\cdot(b+c)=a\cdot b+a\cdot c$

Classwork

Fluency Exercise (6 minutes): Integer Addition

Photocopy the attached 2-page fluency-building exercises so that each student receives a copy. Before students begin, inform them that they may not skip over questions and that they must move in order. Time students, allowing one minute to complete Side A. After one minute, discuss the answers. Before administering Side B, elicit strategies from those students who were able to accurately complete many problems on Side A. Administer Side B in the same fashion, and review the answers. Refer to the Sprints and Sprint Delivery Script sections in the Module Overview for directions to administer a Sprint.

Opening Exercise (3 minutes): Recall of a Number's Opposite

This warm-up prepares students for Exercise 1. Ahead of time, post a large number line on the side wall (either in poster form or with painter's tape).

> *Scaffolding:*
>
> Select specific cards to give to students to challenge them at their level.

As students enter the room, hand them a small sticky note with a rational number on it. Ask them to "find their opposite." (Sticky notes are such that each signed number has a "match" for opposite.) Students pair up according to opposites, walk to the number line on the side wall, and stick their numbers in the correct locations on the number line. The class comes to a consensus that all numbers are placed in the correct location.

Example 1 (4 minutes): The Opposite of a Sum is the Sum of its Opposites

Have the following statement up on the board: "The opposite of a sum is the sum of its opposites." Tell students we are going to use some numbers from the Opening Exercise to investigate this statement.

Ask two pairs of students (who were partners from the Opening Exercise) to come to the front of the room. (Choose students who had rational numbers that were integers, as they will be easier to understand in this example.) Have one person from each pair write their numbers on the board; let's say they were 7 and -2. Find the sum, $7 + -2 = 5$, and then find the opposite of the sum, -5. Now have partners write their numbers on the board, -7 and 2, and then find the sum of these opposites, -5. Now it is clear that the opposite of the sum is equal to the sum of the opposites.

Example 1: The Opposite of a Sum is the Sum of its Opposites

Explain the meaning of "The opposite of a sum is the sum of its opposites." Use a specific math example.

Rational Number	Rational Number	Sum	Opposite of the Sum
7	-2	5	-5

Opposite Rational Number	Opposite Rational Number	Sum
-7	2	-5

It means that if you have a sum and want to take the opposite, for instance, $-(7 + (-2))$, you can rewrite it as the sum of each addend's opposite, $-7 + 2$.

> *Scaffolding:*
>
> Display an anchor poster in the classroom to show the meaning of "the opposite of a sum is the sum of its opposites." Label the "opposite" and "sum" in a specific math example.

Exercise 1 (5 minutes)

Have students arrive at an answer to the following. Students share their different strategies with the class. The class members discuss the strategies they used. They determine which are most efficient, which ways are less likely to cause errors and confusion, and whether they were able to reach the correct answer. If no students share the solution method on the right, share it with the class.

Exercise 1

Represent the following expression with a single rational number.

$$-2\frac{2}{5} + 3\frac{1}{4} - \frac{3}{5}$$

Two Possible Methods:

$-2\frac{8}{20} + 3\frac{5}{20} - \frac{12}{20}$

$-\frac{48}{20} + \frac{65}{20} - \frac{12}{20}$

$\frac{17}{20} - \frac{12}{20}$

$\frac{5}{20}$ *or* $\frac{1}{4}$

OR

$-2\frac{2}{5} + 3\frac{1}{4} + \left(-\frac{3}{5}\right)$

$-2\frac{2}{5} + \left(-\frac{3}{5}\right) + 3\frac{1}{4}$ *commutative property*

$-2\frac{5}{5} + 3\frac{1}{4}$

$-3 + 3\frac{1}{4}$

$\frac{1}{4}$

After students share their strategies, the following are questions that may guide the whole-group discussion. Student responses to the suggested discussion questions will vary.

- Was it difficult for you to add the mixed numbers with different signs and denominators? Why or why not?
- Were you able to arrive at the correct answer?
- Which method do you prefer?
- Which method is more challenging for you?

Example 2 (5 minutes): A Mixed Number Is a Sum

The following example allows students to focus on a mixed number as a sum. Looking at $2\frac{2}{5}$, they think about how it can be rewritten using addition. (It means $2 + \frac{2}{5}$.) Once students represent it as a sum, they recognize that $-2\frac{2}{5}$ means $-2 + \left(-\frac{2}{5}\right)$. The following is a possible lead-in question.

- $-2\frac{2}{5}$ is the opposite of $2\frac{2}{5}$. How can we show "the opposite of a sum is the sum of its opposites" with the number $-2\frac{2}{5}$? How do we model it on a number line?

Example 2: A Mixed Number Is a Sum

Use the number line model shown below to explain and write the opposite of $2\frac{2}{5}$ as a sum of two rational numbers.

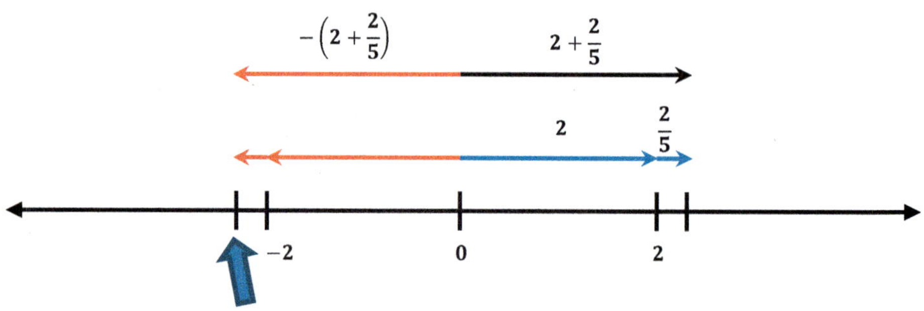

The opposite of a sum (top single arrow pointing left) and the sum of the opposites correspond to the same point on the number line.

The opposite of $2\frac{2}{5}$ is $-2\frac{2}{5}$.

$-2\frac{2}{5}$ written as the sum of two rational numbers is $-2 + \left(-\frac{2}{5}\right)$.

Exercise 2 (2 minutes)

Students independently rewrite each mixed number as the sum of two signed numbers. The teacher circulates the room, providing assistance as needed. After two minutes, discuss the answers as a whole group.

> *Scaffolding:*
> Provide students with a laminated copy of the number line model used in Example 2. Also provide number lines so they can represent each of the following as a sum on the number line.

Exercise 2

Rewrite each mixed number as the sum of two signed numbers.

a. $-9\frac{5}{8}$

$-9+\left(-\frac{5}{8}\right)$

b. $-2\frac{1}{2}$

$-2+\left(-\frac{1}{2}\right)$

c. $8\frac{11}{12}$

$8+\frac{11}{12}$

Exercise 3 (2 minutes)

MP.8 Students independently use the reverse process to represent each sum or difference as a mixed number. The teacher circulates the room providing assistance as needed. After two minutes, discuss the answers as a whole group.

Exercise 3

Represent each sum as a mixed number.

a. $-1+\left(-\frac{5}{12}\right)$

$-1\frac{5}{12}$

b. $30+\frac{1}{8}$

$30\frac{1}{8}$

c. $-17+\left(-\frac{1}{9}\right)$

$-17\frac{1}{9}$

Note: Exercises 3 and 4 are designed to provide students with an opportunity to practice writing mixed numbers as sums so they can do so as the need arises in more complicated problems.

Exercise 4 (5 minutes)

Students work independently to solve the problem below. Then student volunteers share their steps and solutions with the class. Note, the solution below includes just one possible solution method. However, a common mistake is for students to arrive at an *incorrect answer of* $-5\frac{1}{8}$. As needed, revisit subtracting a mixed number from a whole number.

Exercise 4

Mr. Mitchell lost 10 pounds over the summer by jogging each week. By winter, he had gained $5\frac{1}{8}$ pounds. Represent this situation with an expression involving signed numbers. What is the overall change in Mr. Mitchell's weight?

$-10+5\frac{1}{8}$

$=-10+5+\frac{1}{8}$

$=(-10+5)+\frac{1}{8}$

$=(-5)+\frac{1}{8}$

$=-4\frac{7}{8}$

Mr. Mitchell's weight dropped by $4\frac{7}{8}$ pounds.

Exercise 5 (7 minutes)

MP.1

Students work with a partner to complete the following exercise. Students make sense of each step and come up with an alternate method of solving the problem.

After five minutes, class resumes as a whole group, and students volunteer verbal explanations and their own methods for solving the problem.

Exercise 5

Jamal is completing a math problem and represents the expression $-5\frac{5}{7}+8-3\frac{2}{7}$ with a single rational number as shown in the steps below. Justify each of Jamal's steps. Then, show another way to solve the problem.

$= -5\frac{5}{7}+8+\left(-3\frac{2}{7}\right)$

$= -5\frac{5}{7}+\left(-3\frac{2}{7}\right)+8$

$= -5+\left(-\frac{5}{7}\right)+(-3)+\left(-\frac{2}{7}\right)+8$

$= -5+\left(-\frac{5}{7}\right)+\left(-\frac{2}{7}\right)+(-3)+8$

$= -5+(-1)+(-3)+8$

$= -6+(-3)+8$

$= (-9)+8$

$= -1$

Scaffolding:

Provide a list of the properties of operations as a reference sheet or as a bookmark for students.

Step 1: *Subtracting a number is the same as adding its inverse.*

Step 2: *Apply the commutative property of addition.*

Step 3: *The opposite of a sum is the sum of its opposites.*

Step 4: *Apply the commutative property of addition.*

Step 5: *Apply the associative property of addition.*

$\left(-\frac{5}{7}\right)+\left(-\frac{2}{7}\right)=\left(-\frac{7}{7}\right)= -1$

Step 6: $-5 + (-1) = -6$

Step 7: $-6 + -3 = -9$

Step 8: $-9 + 8 = -1$

Answers will vary for other methods of reaching a single rational number. Students may choose to add $-5\frac{5}{7}$ and 8 together first, but a common mistake is to represent their sum as $3\frac{5}{7}$, rather than $2\frac{2}{7}$.

Closing (2 minutes)

- How can we rewrite the opposite of a sum?
  - *As the sum of its opposites*
- How is it helpful when finding the sums and differences of rational numbers to use the properties of operations?
  - *It allows us to regroup terms so that we can efficiently arrive at an answer. For instance, in an expression we may wish to first combine certain rational numbers that are in decimal form or those that are in fractional form. Or, we may wish to group together all the negative numbers if we are finding the sum of positive and negative numbers.*

> Lesson Summary
>
> - **Use the properties of operations to add and subtract rational numbers more efficiently. For instance,**
>
> $-5\frac{2}{9}+3.7+5\frac{2}{9}=\left(-5\frac{2}{9}+5\frac{2}{9}\right)+3.7=0+3.7=3.7.$
> - **The opposite of a sum is the sum of its opposites as shown in the examples that follow:**
>
> $-4\frac{4}{7}=-4+\left(-\frac{4}{7}\right).$
>
> $-(5+3)=-5+(-3).$

Exit Ticket (4 minutes)

Name ______________________________ Date ______________

Lesson 8: Applying the Properties of Operations to Add and Subtract Rational Numbers

Exit Ticket

Mariah and Shane both started to work on a math problem and were comparing their work in math class. Are both of their representations correct? Explain, and finish the math problem correctly to arrive at the correct answer.

Math Problem
Jessica's friend lent her $5. Later that day Jessica gave her friend back $1\frac{3}{4}$ dollars. Which rational number represents the overall change to the amount of money Jessica's friend has?

Mariah started the problem as follows:

$$-5 - \left(-1\frac{3}{4}\right)$$
$$= -5 + 1 - \frac{3}{4}$$

Shane started the problem as follows:

$$-5 - \left(-1\frac{3}{4}\right)$$
$$= -5 + \left(1\frac{3}{4}\right)$$
$$= -5 + \left(1 + \frac{3}{4}\right)$$

____

____

____

____

____

Exit Ticket Sample Solutions

Mariah and Shane both started to work on a math problem and were comparing their work in math class. Are both of their representations correct? Explain, and finish the math problem correctly to arrive at the correct answer.

Math Problem

Jessica's friend lent her $\$5$. Later that day Jessica gave her friend back $1\frac{3}{4}$ dollars.

Which rational number represents the overall change to the amount of money Jessica's friend has?

Mariah started the problem as follows:

$$-5-\left(-1\frac{3}{4}\right)$$
$$= -5 + 1 - \frac{3}{4}$$

Shane started the problem as follows:

$$-5-\left(-1\frac{3}{4}\right)$$
$$= -5+\left(1\frac{3}{4}\right)$$
$$= -5+\left(1+\frac{3}{4}\right)$$

Shane's method is correct. In Mariah's math work, she only dealt with part of the mixed number. The fractional part should have been positive too because the opposite of $-1\frac{3}{4}$ is $1\frac{3}{4}$, which contains both a positive 1 and a positive $\frac{3}{4}$. The correct work would be

$$-5-\left(-1\frac{3}{4}\right) = -5+\left(1\frac{3}{4}\right) = -5+\left(1+\frac{3}{4}\right) = (-5+1)+\frac{3}{4} = -4+\frac{3}{4} = -3\frac{1}{4}.$$

The rational number would be $-3\frac{1}{4}$, which means Jessica's friend gave away $3\frac{1}{4}$ dollars, or $\$3.25$.

Problem Set Sample Solutions

1. **Represent each sum as a single rational number.**

 a. $-14+\left(-\frac{8}{9}\right)$

 $-14\frac{8}{9}$

 b. $7+\frac{1}{9}$

 $7\frac{1}{9}$

 c. $-3+\left(-\frac{1}{6}\right)$

 $-3\frac{1}{6}$

Rewrite each of the following to show that *the opposite of a sum is the sum of the opposites*. Problem 2 has been completed as an example.

2. $-(9+8) = -9+(-8)$

 $-17 = -17$ *Answer provided in student materials.*

3. $-\left(\frac{1}{4}+6\right)=-\frac{1}{4}+(-6)$

$-6\frac{1}{4}=-6\frac{1}{4}$

4. $-(10+(-6))=-10+6$

$-4=-4$

5. $-\left((-55)+\frac{1}{2}\right)=55+\left(-\frac{1}{2}\right)$

$54\frac{1}{2}=54\frac{1}{2}$

Use your knowledge of rational numbers to answer the following questions.

6. **Meghan said the opposite of the sum of -12 and 4 is 8. Do you agree? Why or why not?**

Yes, I agree. The sum of -12 and 4 is -8, and the opposite of -8 is 8.

7. **Jolene lost her wallet at the mall. It had $\$10$ in it. When she got home, her brother felt sorry for her and gave her $\$5.75$. Represent this situation with an expression involving rational numbers. What is the overall change in the amount of money Jolene has?**

$-10+5.75=-4.25$. *The overall change in the amount of money Jolene has is -4.25 dollars.*

8. **Isaiah is completing a math problem and is at the last step: $25-28\frac{1}{5}$. What is the answer? Show your work.**

$25-28\frac{1}{5}=25+\left(-28+\left(-\frac{1}{5}\right)\right)=(25+-28)+\left(-\frac{1}{5}\right)=-3\frac{1}{5}$

9. **A number added to its opposite equals zero. What do you suppose is true about *a sum added to its opposite*?**

Use the following examples to reach a conclusion. Express the answer to each example as a single rational number.

A sum added to its opposite is zero.

a. $(3+4)+(-3+-4)=7+(-7)=0$

b. $(-8+1)+(8+(-1))=(-7)+7=0$

c. $\left(-\frac{1}{2}+\left(-\frac{1}{4}\right)\right)+\left(\frac{1}{2}+\frac{1}{4}\right)=\left(-\frac{3}{4}\right)+\frac{3}{4}=0$

Number Correct: ______

Integer Addition—Round 1

Directions: Determine the sum of the integers, and write it in the column to the right.

1.	$8+(-5)$		18.	$-38+25$	
2.	$10+(-3)$		19.	$-19+(-11)$	
3.	$2+(-7)$		20.	$2+(-7)$	
4.	$4+(-11)$		21.	$-23+(-23)$	
5.	$-3+(-9)$		22.	$45+(-32)$	
6.	$-12+(-7)$		23.	$16+(-24)$	
7.	$-13+5$		24.	$-28+13$	
8.	$-4+9$		25.	$-15+15$	
9.	$7+(-7)$		26.	$12+(-19)$	
10.	$-13+13$		27.	$-24+(-32)$	
11.	$14+(-20)$		28.	$-18+(-18)$	
12.	$6+(-4)$		29.	$14+(-26)$	
13.	$10+(-7)$		30.	$-7+8+(-3)$	
14.	$-16+9$		31.	$2+(-15)+4$	
15.	$-10+34$		32.	$-8+(-19)+(-11)$	
16.	$-20+(-5)$		33.	$15+(-12)+7$	
17.	$-18+15$		34.	$-28+7+(-7)$	

Integer Addition—Round 1 [KEY]

Directions: Determine the sum of the integers, and write it in the column to the right.

1.	$8 + (-5)$	3	18.	$-38 + 25$	-13
2.	$10 + (-3)$	7	19.	$-19 + (-11)$	-30
3.	$2 + (-7)$	-5	20.	$2 + (-7)$	-5
4.	$4 + (-11)$	-7	21.	$-23 + (-23)$	-46
5.	$-3 + (-9)$	-12	22.	$45 + (-32)$	13
6.	$-12 + (-7)$	-19	23.	$16 + (-24)$	-8
7.	$-13 + 5$	-8	24.	$-28 + 13$	-15
8.	$-4 + 9$	5	25.	$-15 + 15$	0
9.	$7 + (-7)$	0	26.	$12 + (-19)$	-7
10.	$-13 + 13$	0	27.	$-24 + (-32)$	-56
11.	$14 + (-20)$	-6	28.	$-18 + (-18)$	-36
12.	$6 + (-4)$	2	29.	$14 + (-26)$	-12
13.	$10 + (-7)$	3	30.	$-7 + 8 + (-3)$	-2
14.	$-16 + 9$	-7	31.	$2 + (-15) + 4$	-9
15.	$-10 + 34$	24	32.	$-8 + (-19) + (-11)$	-38
16.	$-20 + (-5)$	-25	33.	$15 + (-12) + 7$	10
17.	$-18 + 15$	-3	34.	$-28 + 7 + (-7)$	-28

Number Correct: ______

Improvement: ______

Integer Addition—Round 2

Directions: Determine the sum of the integers, and write it in the column to the right.

1.	$5 + (-12)$	
2.	$10 + (-6)$	
3.	$-9 + (-13)$	
4.	$-12 + 17$	
5.	$-15 + 15$	
6.	$16 + (-25)$	
7.	$-12 + (-8)$	
8.	$-25 + (-29)$	
9.	$28 + (-12)$	
10.	$-19 + (-19)$	
11.	$-17 + 20$	
12.	$8 + (-18)$	
13.	$13 + (-15)$	
14.	$-10 + (-16)$	
15.	$35 + (-35)$	
16.	$9 + (-14)$	
17.	$-34 + (-27)$	

18.	$23 + (-31)$	
19.	$-26 + (-19)$	
20.	$16 + (-37)$	
21.	$-21 + 14$	
22.	$33 + (-8)$	
23.	$-31 + (-13)$	
24.	$-16 + 16$	
25.	$30 + (-43)$	
26.	$-22 + (-18)$	
27.	$-43 + 27$	
28.	$38 + (-19)$	
29.	$-13 + (-13)$	
30.	$5 + (-8) + (-3)$	
31.	$6 + (-11) + 14$	
32.	$-17 + 5 + 19$	
33.	$-16 + (-4) + (-7)$	
34.	$8 + (-24) + 12$	

Integer Addition—Round 2 [KEY]

Directions: Determine the sum of the integers, and write it in the column to the right.

1.	$5+(-12)$	-7	18.	$23+(-31)$	-8
2.	$10+(-6)$	4	19.	$-26+(-19)$	-45
3.	$-9+(-13)$	-22	20.	$16+(-37)$	-21
4.	$-12+17$	5	21.	$-21+14$	-7
5.	$-15+15$	0	22.	$33+(-8)$	25
6.	$16+(-25)$	-9	23.	$-31+(-13)$	-44
7.	$-12+(-8)$	-20	24.	$-16+16$	0
8.	$-25+(-29)$	-54	25.	$30+(-43)$	-13
9.	$28+(-12)$	16	26.	$-22+(-18)$	-40
10.	$-19+(-19)$	-38	27.	$-43+27$	-16
11.	$-17+20$	3	28.	$38+(-19)$	19
12.	$8+(-18)$	-10	29.	$-13+(-13)$	-26
13.	$13+(-15)$	-2	30.	$5+(-8)+(-3)$	-6
14.	$-10+(-16)$	-26	31.	$6+(-11)+14$	9
15.	$35+(-35)$	0	32.	$-17+5+19$	7
16.	$9+(-14)$	-5	33.	$-16+(-4)+(-7)$	-27
17.	$-34+(-27)$	-61	34.	$8+(-24)+12$	-4

Lesson 9: Applying the Properties of Operations to Add and Subtract Rational Numbers

Student Outcomes

- Students use properties of operations to add and subtract rational numbers without the use of a calculator.
- Students recognize that any problem involving addition and subtraction of rational numbers can be written as a problem using addition and subtraction of positive numbers only.
- Students use the commutative and associative properties of addition to rewrite numerical expressions in different forms. They know that the opposite of a sum is the sum of the opposites; for example, $-(3-4)=-3+4$.

Classwork

Fluency Exercise (6 minutes): Integer Subtraction

Photocopy the attached 2-page fluency-building exercises so that each student receives a copy. Before students begin, inform them that they may not skip over questions and that they must move in order. Time students, allowing one minute to complete Side A. After one minute, discuss the answers. Before administering Side B, elicit strategies from those students who were able to accurately complete many problems on Side A. Administer Side B in the same fashion, and review the answers. Refer to the Sprints and Sprint Delivery Script sections in the Module Overview for directions to administer a Sprint.

Exercise 1 (6 minutes)

MP.8 Students are given scrambled steps to one possible solution to the following problem.[1] They work independently to arrange the expressions in an order that leads to a solution and record their solutions in the student materials.

Exercise 1

Unscramble the cards, and show the steps in the correct order to arrive at the solution to $5\frac{2}{9}-\left(8.1+5\frac{2}{9}\right)$.

$0+(-8.1)$	$\left(5\frac{2}{9}+\left(-5\frac{2}{9}\right)\right)+(-8.1)$	-8.1
$5\frac{2}{9}+\left(-8.1+\left(-5\frac{2}{9}\right)\right)$	$5\frac{2}{9}+\left(-5\frac{2}{9}+(-8.1)\right)$	

[1]The scrambled steps may also be displayed on an interactive white board, and students can come up one at a time to slide a step into the correct position.

G7-M2-B2-v1.3.0_05/04/15

$5\frac{2}{9}+\left(-8.1+\left(-5\frac{2}{9}\right)\right)$	*The opposite of a sum is the sum of its opposites.*
$5\frac{2}{9}+\left(-5\frac{2}{9}+(-8.1)\right)$	*Apply the commutative property of addition.*
$\left(5\frac{2}{9}+\left(-5\frac{2}{9}\right)\right)+(-8.1)$	*Apply the associative property of addition.*
$0+(-8.1)$	*A number plus its opposite equals zero.*
-8.1	*Apply the additive identity property.*

Scaffolding:
- Adapt for struggling learners by having fewer steps to rearrange or by including only integers.
- Adapt for proficient students by requiring that they state the property of operations to justify each step.

After 2 minutes, students share the correct sequence of steps with the class.

- What allows us to represent operations in another form and rearrange the order of terms?
  - *The properties of operations*
- Specifically, which properties of operations were used in this example?
  - *Students recall the additive inverse property and commutative property of addition. (Students are reminded to focus on all of the properties that justify their steps today.)*
- Why did we use the properties of operations?
  - *Students recognize that using the properties allows us to efficiently (more easily) calculate the answer to the problem.*

Examples 1–2 (7 minutes)

Students record the following examples. Students assist in volunteering verbal explanations for each step during the whole-group discussion. Today, students' focus is *not* on memorizing the names of each property but rather on knowing that each representation is justifiable through the properties of operations.

Scaffolding:

Provide a list of the properties of operations as a reference sheet for students.

Examples 1–2

Represent each of the following expressions as one rational number. Show and explain your steps.

1. $4\frac{4}{7}-\left(4\frac{4}{7}-10\right)$

$=4\frac{4}{7}-\left(4\frac{4}{7}+(-10)\right)$	*Subtracting a number is the same as adding its inverse.*
$=4\frac{4}{7}+\left(-4\frac{4}{7}+10\right)$	*The opposite of a sum is the sum of its opposites.*
$=\left(4\frac{4}{7}+\left(-4\frac{4}{7}\right)\right)+10$	*The associative property of addition*
$=0+10$	*A number plus its opposite equals zero.*
$=10$	

- First, predict the answer. Explain your prediction.
  - *Answers may vary.*

EUREKA MATH™

- The answer will be between 0 and $\frac{1}{2}$ because $5 + (-5) = 0$ and $-4\frac{4}{7}$ is close to -5, but 5 has a larger absolute value than $-4\frac{4}{7}$. To add $5 + \left(-4\frac{4}{7}\right)$, we subtract their absolute values. Since $-4\frac{4}{7}$ is close to $-4\frac{1}{2}$, the answer will be about $5 - 4\frac{1}{2} = \frac{1}{2}$.

2. $5 + \left(-4\frac{4}{7}\right)$

$= 5 + \left(-\left(4 + \frac{4}{7}\right)\right)$	*The mixed number $4\frac{4}{7}$ is equivalent to $4 + \frac{4}{7}$.*
$= 5 + \left(-4 + \left(-\frac{4}{7}\right)\right)$	*The opposite of a sum is the sum of its opposites.*
$= (5 + (-4)) + \left(-\frac{4}{7}\right)$	*Associative property of addition*
$= 1 + \left(-\frac{4}{7}\right)$	$5 + (-4) = 1$
$= \frac{7}{7} + \left(-\frac{4}{7}\right)$	$\frac{7}{7} = 1$
$= \frac{3}{7}$	

- Does our answer match our prediction?
  - *Yes, we predicted a positive number close to zero.*

Exercise 2 (10 minutes): Team Work!

Students work in groups of three. Each student has a different colored pencil. Each problem has at least three steps. Students take turns writing a step to each problem, passing the paper to the next person, and rotating the student who starts first with each new problem.

MP.2 & MP.3

After 8 minutes, students partner up with another group of students to discuss or debate their answers. Students should also explain their steps and the properties/rules that justify each step.

Exercise 2: Team Work!

a. $-5.2 - (-3.1) + 5.2$

$= -5.2 + 3.1 + 5.2$

$= -5.2 + 5.2 + 3.1$

$= 0 + 3.1$

$= 3.1$

b. $32 + \left(-12\frac{7}{8}\right)$

$= 32 + \left(-12 + \left(-\frac{7}{8}\right)\right)$

$= (32 + (-12)) + \left(-\frac{7}{8}\right)$

$= 20 + \left(-\frac{7}{8}\right)$

$= 19\frac{1}{8}$

c. $3\frac{1}{6} + 20.3 - \left(-5\frac{5}{6}\right)$

$= 3\frac{1}{6} + 20.3 + 5\frac{5}{6}$

$= 3\frac{1}{6} + 5\frac{5}{6} + 20.3$

$= 8\frac{6}{6} + 20.3$

$= 9 + 20.3$

$= 29.3$

d. $\frac{16}{20} - (-1.8) - \frac{4}{5}$

$= \frac{16}{20} + 1.8 - \frac{4}{5}$

$= \frac{16}{20} + 1.8 + \left(-\frac{4}{5}\right)$

$= \frac{16}{20} + \left(-\frac{4}{5}\right) + 1.8$

$= \frac{16}{20} + \left(-\frac{16}{20}\right) + 1.8$

$= 0 + 1.8$

$= 1.8$

Exercise 3 (5 minutes)

Students work independently to answer the following question. Then, after 3 minutes, group members share their responses with one another and come to a consensus.

Exercise 3

Explain, step by step, how to arrive at a single rational number to represent the following expression. Show both a written explanation and the related math work for each step.

$$-24 - \left(-\frac{1}{2}\right) - 12.5$$

Subtracting $\left(-\frac{1}{2}\right)$ *is the same as adding its inverse* $\frac{1}{2}$*:* $= -24 + \frac{1}{2} + (-12.5)$

Next, I used the commutative property of addition to rewrite the expression: $= -24 + (-12.5) + \frac{1}{2}$

Next, I added both negative numbers: $= -36.5 + \frac{1}{2}$

Next, I wrote $\frac{1}{2}$ *in its decimal form:* $= -36.5 + 0.5$

Lastly, I added $-36.5 + 0.5$*:* $= -36$

Closing (3 minutes)

- How are the properties of operations helpful when finding the sums and differences of rational numbers?
  - *The properties of operations allow us to add and subtract rational numbers more efficiently.*
- Do you think the properties of operations could be used in a similar way to aid in the multiplication and division of rational numbers?
  - *Answers will vary.*

Lesson Summary

- **Use the properties of operations to add and subtract rational numbers more efficiently. For instance,**

  $-5\frac{2}{9} + 3.7 + 5\frac{2}{9} = \left(-5\frac{2}{9} + 5\frac{2}{9}\right) + 3.7 = 0 + 3.7 = 3.7.$

- **The opposite of a sum is the sum of its opposites as shown in the examples that follow:**

  $-4\frac{4}{7} = -4 + \left(-\frac{4}{7}\right).$

  $-(5 + 3) = -5 + (-3).$

Exit Ticket (8 minutes)

G7-M2-B2-v1.3.0_05/04/15

Name ______________________________ Date ______________

Lesson 9: Applying the Properties of Operations to Add and Subtract Rational Numbers

Exit Ticket

1. Jamie was working on his math homework with his friend, Kent. Jamie looked at the following problem.

$$-9.5 - (-8) - 6.5$$

He told Kent that he did not know how to subtract negative numbers. Kent said that he knew how to solve the problem using only addition. What did Kent mean by that? Explain. Then, show your work, and represent the answer as a single rational number.

____

____

____

Work Space:

Answer: ________________

2. Use one rational number to represent the following expression. Show your work.

$3 + (-0.2) - 15\frac{1}{4}$

Exit Ticket Sample Solutions

1. **Jamie was working on his math homework with his friend, Kent. Jamie looked at the following problem.**

$$-9.5 - (-8) - 6.5$$

He told Kent that he did not know how to subtract negative numbers. Kent said that he knew how to solve the problem using only addition. What did Kent mean by that? Explain. Then, show your work, and represent the answer as a single rational number.

Kent meant that since any subtraction problem can be written as an addition problem by adding the opposite of the number you are subtracting, Jamie can solve the problem by using only addition.

Work Space:

$$-9.5 - (-8) - 6.5$$
$$= -9.5 + 8 + (-6.5)$$
$$= -9.5 + (-6.5) + 8$$
$$= -16 + 8$$
$$= -8$$

Answer: -8

2. **Use one rational number to represent the following expression. Show your work.**

$3 + (-0.2) - 15\frac{1}{4}$

$= 3 + (-0.2) + \left(-15 + \left(-\frac{1}{4}\right)\right)$

$= 3 + \left(-0.2 + (-15) + (-0.25)\right)$

$= 3 + (-15.45)$

$= -12.45$

Problem Set Sample Solutions

Show all steps taken to rewrite each of the following as a single rational number.

1. $80 + \left(-22\frac{4}{15}\right)$

$= 80 + \left(-22 + \left(-\frac{4}{15}\right)\right)$

$= \left(80 + (-22)\right) + \left(-\frac{4}{15}\right)$

$= 58 + \left(-\frac{4}{15}\right)$

$= 57\frac{11}{15}$

2. $10 + \left(-3\frac{3}{8}\right)$

$= 10 + \left(-3 + \left(-\frac{3}{8}\right)\right)$

$= \left(10 + (-3)\right) + \left(-\frac{3}{8}\right)$

$= 7 + \left(-\frac{3}{8}\right)$

$= 6\frac{5}{8}$

3. $\frac{1}{5} + 20.3 - \left(-5\frac{3}{5}\right)$

$= \frac{1}{5} + 20.3 + 5\frac{3}{5}$

$= \frac{1}{5} + 5\frac{3}{5} + 20.3$

$= 5\frac{4}{5} + 20.3$

$= 5\frac{4}{5} + 20\frac{3}{10}$

$= 5\frac{8}{10} + 20\frac{3}{10}$

$= 25\frac{11}{10}$

$= 26\frac{1}{10}$

4. $\frac{11}{12} - (-10) - \frac{5}{6}$

$= \frac{11}{12} + 10 + \left(-\frac{5}{6}\right)$

$= \frac{11}{12} + \left(-\frac{5}{6}\right) + 10$

$= \frac{11}{12} + \left(-\frac{10}{12}\right) + 10$

$= \frac{1}{12} + 10$

$= 10\frac{1}{12}$

5. **Explain, step by step, how to arrive at a single rational number to represent the following expression. Show both a written explanation and the related math work for each step.**

$$1 - \frac{3}{4} + \left(-12\frac{1}{4}\right)$$

First, I rewrote the subtraction of $\frac{3}{4}$ *as the addition of its inverse* $-\frac{3}{4}$*:* $= 1 + \left(-\frac{3}{4}\right) + \left(-12\frac{1}{4}\right)$

Next, I used the associative property of addition to regroup the addend: $= 1 + \left(\left(-\frac{3}{4}\right) + \left(-12\frac{1}{4}\right)\right)$

Next, I separated $-12\frac{1}{4}$ *into the sum of* -12 *and* $-\frac{1}{4}$*:* $= 1 + \left(\left(-\frac{3}{4}\right) + (-12) + \left(-\frac{1}{4}\right)\right)$

Next, I used the commutative property of addition: $= 1 + \left(\left(-\frac{3}{4}\right) + \left(-\frac{1}{4}\right) + (-12)\right)$

Next, I found the sum of $-\frac{3}{4}$ *and* $-\frac{1}{4}$*:* $= 1 + ((-1) + (-12))$

Next, I found the sum of -1 *and* -12*:* $= 1 + (-13)$

Lastly, since the absolute value of 13 *is greater than the absolute value of* 1*, and it is a negative* 13*, the answer will be a negative number. The absolute value of* 13 *minus the absolute value of* 1 *equals* 12*, so the answer is* -12*.* $= -12$

Number Correct: ______

Integer Subtraction—Round 1

Directions: Determine the difference of the integers, and write it in the column to the right.

1.	$4 - 2$		23.	$(-6) - 5$	
2.	$4 - 3$		24.	$(-6) - 7$	
3.	$4 - 4$		25.	$(-6) - 9$	
4.	$4 - 5$		26.	$(-14) - 9$	
5.	$4 - 6$		27.	$(-25) - 9$	
6.	$4 - 9$		28.	$(-12) - 12$	
7.	$4 - 10$		29.	$(-26) - 26$	
8.	$4 - 20$		30.	$(-13) - 21$	
9.	$4 - 80$		31.	$(-25) - 75$	
10.	$4 - 100$		32.	$(-411) - 811$	
11.	$4 - (-1)$		33.	$(-234) - 543$	
12.	$4 - (-2)$		34.	$(-3) - (-1)$	
13.	$4 - (-3)$		35.	$(-3) - (-2)$	
14.	$4 - (-7)$		36.	$(-3) - (-3)$	
15.	$4 - (-17)$		37.	$(-3) - (-4)$	
16.	$4 - (-27)$		38.	$(-3) - (-8)$	
17.	$4 - (-127)$		39.	$(-30) - (-45)$	
18.	$14 - (-6)$		40.	$(-27) - (-13)$	
19.	$23 - (-8)$		41.	$(-13) - (-27)$	
20.	$8 - (-23)$		42.	$(-4) - (-3)$	
21.	$51 - (-3)$		43.	$(-3) - (-4)$	
22.	$48 - (-5)$		44.	$(-1{,}066) - (-34)$	

Integer Subtraction—Round 1 [KEY]

Directions: Determine the difference of the integers, and write it in the column to the right.

1.	$4-2$	2	23.	$(-6)-5$	-11
2.	$4-3$	1	24.	$(-6)-7$	-13
3.	$4-4$	0	25.	$(-6)-9$	-15
4.	$4-5$	-1	26.	$(-14)-9$	-23
5.	$4-6$	-2	27.	$(-25)-9$	-34
6.	$4-9$	-5	28.	$(-12)-12$	-24
7.	$4-10$	-6	29.	$(-26)-26$	-52
8.	$4-20$	-16	30.	$(-13)-21$	-34
9.	$4-80$	-76	31.	$(-25)-75$	-100
10.	$4-100$	-96	32.	$(-411)-811$	$-1,222$
11.	$4-(-1)$	5	33.	$(-234)-543$	-777
12.	$4-(-2)$	6	34.	$(-3)-(-1)$	-2
13.	$4-(-3)$	7	35.	$(-3)-(-2)$	-1
14.	$4-(-7)$	11	36.	$(-3)-(-3)$	0
15.	$4-(-17)$	21	37.	$(-3)-(-4)$	1
16.	$4-(-27)$	31	38.	$(-3)-(-8)$	5
17.	$4-(-127)$	131	39.	$(-30)-(-45)$	15
18.	$14-(-6)$	20	40.	$(-27)-(-13)$	-14
19.	$23-(-8)$	31	41.	$(-13)-(-27)$	14
20.	$8-(-23)$	31	42.	$(-4)-(-3)$	-1
21.	$51-(-3)$	54	43.	$(-3)-(-4)$	1
22.	$48-(-5)$	53	44.	$(-1{,}066)-(-34)$	$-1,032$

Number Correct: ______

Improvement: ______

Integer Subtraction—Round 2

Directions: Determine the difference of the integers, and write it in the column to the right.

1.	$3 - 2$	
2.	$3 - 3$	
3.	$3 - 4$	
4.	$3 - 5$	
5.	$3 - 6$	
6.	$3 - 9$	
7.	$3 - 10$	
8.	$3 - 20$	
9.	$3 - 80$	
10.	$3 - 100$	
11.	$3 - (-1)$	
12.	$3 - (-2)$	
13.	$3 - (-3)$	
14.	$3 - (-7)$	
15.	$3 - (-17)$	
16.	$3 - (-27)$	
17.	$3 - (-127)$	
18.	$13 - (-6)$	
19.	$24 - (-8)$	
20.	$5 - (-23)$	
21.	$61 - (-3)$	
22.	$58 - (-5)$	

23.	$(-8) - 5$	
24.	$(-8) - 7$	
25.	$(-8) - 9$	
26.	$(-15) - 9$	
27.	$(-35) - 9$	
28.	$(-22) - 22$	
29.	$(-27) - 27$	
30.	$(-14) - 21$	
31.	$(-22) - 72$	
32.	$(-311) - 611$	
33.	$(-345) - 654$	
34.	$(-2) - (-1)$	
35.	$(-2) - (-2)$	
36.	$(-2) - (-3)$	
37.	$(-2) - (-4)$	
38.	$(-2) - (-8)$	
39.	$(-20) - (-45)$	
40.	$(-24) - (-13)$	
41.	$(-13) - (-24)$	
42.	$(-5) - (-3)$	
43.	$(-3) - (-5)$	
44.	$(-1{,}034) - (-31)$	

Integer Subtraction—Round 2 [KEY]

Directions: Determine the difference of the integers, and write it in the column to the right.

1.	$3 - 2$	1	23.	$(-8) - 5$	-13
2.	$3 - 3$	0	24.	$(-8) - 7$	-15
3.	$3 - 4$	-1	25.	$(-8) - 9$	-17
4.	$3 - 5$	-2	26.	$(-15) - 9$	-24
5.	$3 - 6$	-3	27.	$(-35) - 9$	-44
6.	$3 - 9$	-6	28.	$(-22) - 22$	-44
7.	$3 - 10$	-7	29.	$(-27) - 27$	-54
8.	$3 - 20$	-17	30.	$(-14) - 21$	-35
9.	$3 - 80$	-77	31.	$(-22) - 72$	-94
10.	$3 - 100$	-97	32.	$(-311) - 611$	-922
11.	$3 - (-1)$	4	33.	$(-345) - 654$	-999
12.	$3 - (-2)$	5	34.	$(-2) - (-1)$	-1
13.	$3 - (-3)$	6	35.	$(-2) - (-2)$	0
14.	$3 - (-7)$	10	36.	$(-2) - (-3)$	1
15.	$3 - (-17)$	20	37.	$(-2) - (-4)$	2
16.	$3 - (-27)$	30	38.	$(-2) - (-8)$	6
17.	$3 - (-127)$	130	39.	$(-20) - (-45)$	25
18.	$13 - (-6)$	19	40.	$(-24) - (-13)$	-11
19.	$24 - (-8)$	32	41.	$(-13) - (-24)$	11
20.	$5 - (-23)$	28	42.	$(-5) - (-3)$	-2
21.	$61 - (-3)$	64	43.	$(-3) - (-5)$	2
22.	$58 - (-5)$	63	44.	$(-1{,}034) - (-31)$	$-1{,}003$

Topic B

Multiplication and Division of Integers and Rational Numbers

7.NS.A.2

Focus Standard:	7.NS.A.2	Apply and extend previous understandings of multiplication and division and of fractions to multiply and divide rational numbers. a. Understand that multiplication is extended from fractions to rational numbers by requiring that operations continue to satisfy the properties of operations, particularly the distributive property, leading to products such as $(-1)(-1) = 1$ and the rules for multiplying signed numbers. Interpret products of rational numbers by describing real-world contexts. b. Understand that integers can be divided, provided that the divisor is not zero, and every quotient of integers (with non-zero divisor) is a rational number. If *p* and *q* are integers, then $-(p/q) = (-p)/q = p/(-q)$. Interpret quotients of rational numbers by describing real-world contexts. c. Apply properties of operations as strategies to multiply and divide rational numbers. d. Convert a rational number to a decimal number using long division; know that the decimal form of a rational number terminates in 0s or eventually repeats.
Instructional Days:	7	
Lesson 10:	Understanding Multiplication of Integers (P)[1]	
Lesson 11:	Develop Rules for Multiplying Signed Numbers (P)	
Lesson 12:	Division of Integers (P)	
Lesson 13:	Converting Between Fractions and Decimals Using Equivalent Fractions (P)	
Lesson 14:	Converting Rational Numbers to Decimals Using Long Division (P)	
Lesson 15:	Multiplication and Division of Rational Numbers (P)	
Lesson 16:	Applying the Properties of Operations to Multiply and Divide Rational Numbers (S)	

[1]Lesson Structure Key: **P**-Problem Set Lesson, **M**-Modeling Cycle Lesson, **E**-Exploration Lesson, **S**-Socratic Lesson

In Topic B, students extend their understanding of multiplication and division of whole numbers, decimals, and fractions to find the products and quotients of signed numbers (**7.NS.A.2**). Students begin in Lesson 10 by returning to conceptualization of multiplication as repeated addition. They relate multiplication to the Integer Game. For instance, gaining four -5 cards, or $4(-5)$, is the same as $0 + (-5) + (-5) + (-5) + (-5)$, which is the same as $0 - 5 - 5 - 5 - 5$, or -20. They realize that if a negative card is taken out of their hand multiple times, their score goes up, for example, $(-2)(-6) = 0 - (-6) - (-6) = 0 + 6 + 6 = 12$. In Lesson 11, students draw upon their experiences with the Integer Card Game to justify the rules for multiplication of integers. The additive inverse (**7.NS.A.1c**) and distributive property are used to show that $(-1)(-1) = 1$ (**7.NS.A.2a**).

From earlier grades, students understand division as the process of finding the missing factor of a product (**3.OA.B.6**). In Lesson 12, they use this relationship to justify that the rules for dividing signed numbers are consistent with that of multiplication, provided the divisor is not zero (**7.NS.A.2b**). Students extend the integer rules to include all rational numbers, recognizing that every quotient of two integers is a rational number provided the divisor is not zero.

In Lesson 13, students realize that the context of a word problem often determines whether the answer should be expressed in the fractional or decimal form of a rational number. They draw upon their previous understanding of equivalent fractions, place value, and powers of ten to convert fractions whose denominators are a product of 2's and 5's into decimals. In Lesson 14, students use long division to convert any fraction into a decimal that either terminates in zeros or repeats (**7.NS.A.2d**). Products and quotients continue to be related to the real world. In Lesson 15, students create numerical expressions with rational numbers based on the context of word problems. In Lesson 16, properties of operations are used to rewrite expressions in equivalent forms as students multiply and divide rational numbers efficiently without the aid of a calculator (**7.NS.A.2c**).

Lesson 10: Understanding Multiplication of Integers

Student Outcomes

- Students practice and justify their understanding of multiplication of integers by using the Integer Game. For example, 3×5 corresponds to what happens to your score if you get three 5 cards; $3 \times (-5)$ corresponds to what happens to your score if you get three -5 cards; $(-3) \times 5$ corresponds to what happens to your score if you lose three 5 cards; and $(-3) \times (-5)$ corresponds to what happens to your score if you lose three -5 cards.
- Students explain that multiplying by a positive integer is repeated addition and that adding a number multiple times has the same effect as removing the opposite value the same number of times (e.g., $5 \times 3 = (-5) \times (-3)$ and $5 \times (-3) = (-5) \times 3$).
- Students use the properties and facts of operations to extend multiplication of whole numbers to multiplication of integers.

Classwork

Exercise 1 (4 minutes)

In groups of four, students play one round of the Integer Game using the Integer Game Outline as a reference if needed.

> **Exercise 1: Integer Game Revisited**
>
> **In groups of four, play one round of the Integer Game (see Integer Game outline for directions).**

Example 1 (16 minutes): Product of a Positive Integer and a Negative Integer

The numbers used throughout this example are a sample response. Students' answers depend on the cards they pick.

Part A: Students pick four cards. Instruct students to record the values of their cards on the images in Part A. One of the four card images has a ✭ beneath it. The ✭ is used to indicate which of the four cards to copy (or multiply) in Part B.

> **Example 1: Product of a Positive Integer and a Negative Integer**
>
> **Part A:**

3	-2	-5	4
		☆	

Part B: Instruct students to copy the value of the card with the ✭ beneath it from Part A on each card with a ✭ beneath it in Part B. The three remaining card values from Part A are entered in the three remaining card images in Part B. Students now have a total of six integer cards.

Part B:

3	−2	4	−5	−5	−5
			☆	☆	☆

Use your cards from Part B to answer the questions below.

a. **Write a product that describes the three matching cards.**

$3 \times (-5)$

b. **Write an expression that represents how each of the ★ cards changes your score.**

$(-5) + (-5) + (-5)$

c. **Write an equation that relates these two expressions.**

$3 \times (-5) = (-5) + (-5) + (-5)$

d. **Write an integer that represents the total change to your score by the three ★ cards.**

-15

e. **Write an equation that relates the product and how it affects your score.**

$3 \times (-5) = -15$

Part C: Instruct students to pick three new cards and record the values of their cards on the images in Part C. The teacher chooses the fourth card and instructs the class to place a ★ beneath it to indicate which card will be cloned (multiplied) in Part D.

Part C:

−1	5	−3	4
			☆

Part D: Instruct students to record the value of the card with the ★ beneath it from Part C on each image with a ★ beneath it in Part D. Also, rewrite the values of the three remaining cards on the other three images. Students now have a total of eight integer cards.

Part D:

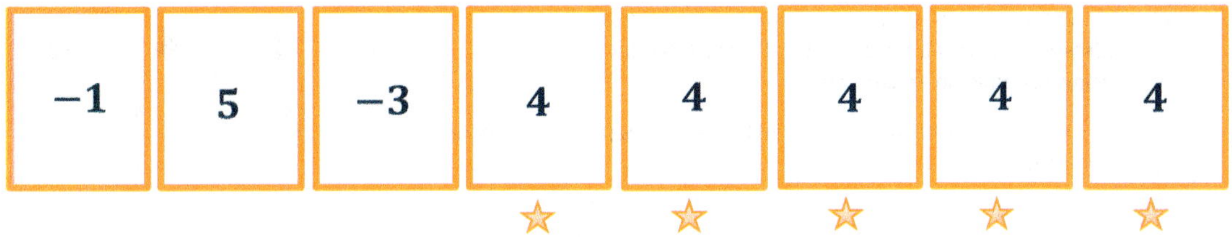

Use your cards from Part D to answer the questions below.

f. Write a product that describes the five matching cards.

5×4

g. Write an expression that represents how each of the ★ cards changes your score.

$4 + 4 + 4 + 4 + 4$

h. Write an equation that relates these two expressions.

$5 \times 4 = 4 + 4 + 4 + 4 + 4$

i. Write an integer that represents the total change to your score by the five ★ cards.

20

j. Write an equation that relates the product and how it affects your score.

$5 \times 4 = 20$

Students write conclusions using their own words in the student materials.

k. Use the expression 5×4 to relate the multiplication of a positive valued card to addition.

Multiplying a positive integer card is repeated addition of the positive integer card and increases your score.

$5 \times 4 = 4 + 4 + 4 + 4 + 4 = 20$

l. Use the expression $3 \times (-5)$ to relate the multiplication of a negative valued card to addition.

Multiplying a negative integer card is repeated addition of the negative integer card and decreases your score.

$3 \times (-5) = (-5) + (-5) + (-5) = -15$

Example 2 (5 minutes): Product of a Negative Integer and a Positive Integer

- If $3 \times (a)$ represents putting three cards with the value a into your playing hand, what would $(-3) \times (a)$ represent?

The student materials provide the sample playing hand from the Integer Game shown below.

Example 2: Product of a Negative Integer and a Positive Integer

a. If all of the 4's from the playing hand on the right are discarded, how will the score be affected? Model this using a product in an equation.

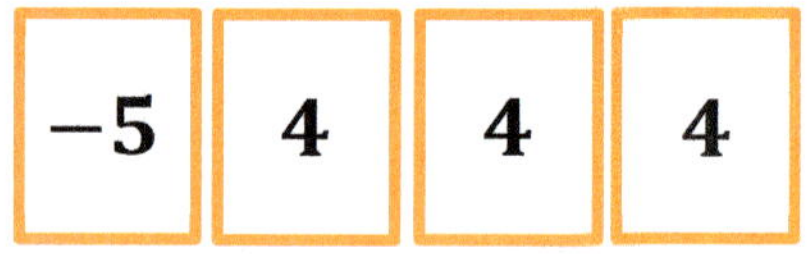

The score decreases by 4 three consecutive times for a total decrease of 12 points. The equation is $-3 \times 4 = -12$.

b. What three matching cards could be added to those pictured to get the same change in score? Model this using a product in an equation.

To get the same change in score, you would add three (-4)'s. The equation is $3 \times (-4) = -12$.

c. Seeing how each play affects the score, relate the products that you used to model them. What do you conclude about multiplying integers with opposite signs?

$(-3) \times 4 = 3 \times (-4)$; adding a value multiple times has the same effect as removing the opposite value the same number of times.

Example 3 (5 minutes): Product of Two Negative Integers

- Using the meaning of $(-3) \times (a)$ from Example 2, what does $(-3) \times (a)$ represent if the value of a is negative?

The student materials provide the sample playing hand from the Integer Game shown below.

Example 3: Product of Two Negative Integers

a. If the matching cards from the playing hand on the right are discarded, how will this hand's score be affected? Model this using a product in an equation.

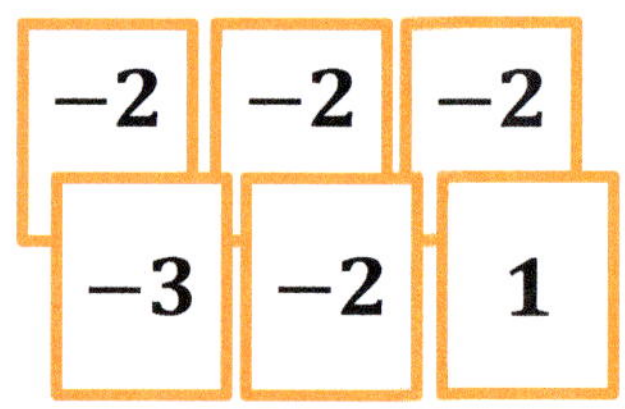

Removing -2 from the set of cards will cause the score to increase by 2. Removing all four of the (-2)'s causes the score to increase by two, four consecutive times for a total increase of 8; $-4 \times (-2) = 8$.

b. What four matching cards could be added to those pictured to get the same change in score? Model this using a product in an equation.

An increase of 8 could come from adding four 2's to the cards shown; $4 \times 2 = 8$.

c. Seeing how each play affects the score, relate the products that you used to model them. What do you conclude about multiplying integers with the same sign?

$-4 \times (-2) = 4 \times 2$; adding a value multiple times has the same effect as removing the opposite value the same number of times.

d. Using the conclusions from Examples 2 and 3, what can we conclude about multiplying integers? Write a few examples.

The product of two integers is equal to the product of their opposites; removing two (-4)'s is the same as adding two 4's; adding three (-5)'s is the same as removing three 5's.

Examples: $(-4) \times (-5) = 4 \times 5$; $(-2) \times 7 = 2 \times (-7)$; $6 \times (-4) = (-6) \times 4$

Removing two (-4)'s is the same as adding two 4's; adding three (-5)'s is the same as removing three 5's.

Closing (5 minutes)

This closing question extends prior knowledge about multiplication of whole numbers as a collection of equal-sized groups onto the family of integers.

- After examining the effects of multiple cards of equal value on scores in the Integer Game, how can we use the representation of 4×5 below to help explain what $4 \times (-5)$ means?

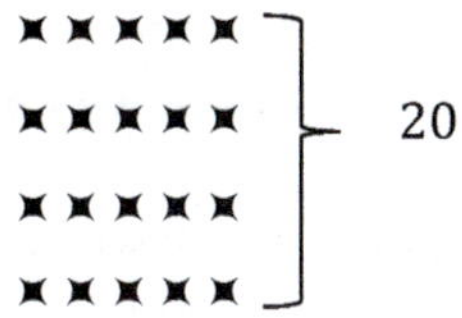

  - *If one row of stars has a value of* (-5)*, then four rows must have a total of* -20.

> **Lesson Summary**
>
> **Multiplying integers is repeated addition and can be modeled with the Integer Game. If $3 \times a$ corresponds to what happens to your score if you get three cards of value a, then $(-3) \times a$ corresponds to what happens to your score if you lose three cards of value a. Adding a number multiple times has the same effect as removing the opposite value the same number of times (e.g., $a \times b = (-a) \times (-b)$ and $a \times (-b) = (-a) \times b$).**

Exit Ticket (10 minutes)

Name ______________________________ Date ______________

Lesson 10: Understanding Multiplication of Integers

Exit Ticket

1. Natalie is playing the Integer Game and only shows you the four cards shown below. She tells you that the rest of her cards have the same values on them and match one of these four cards.

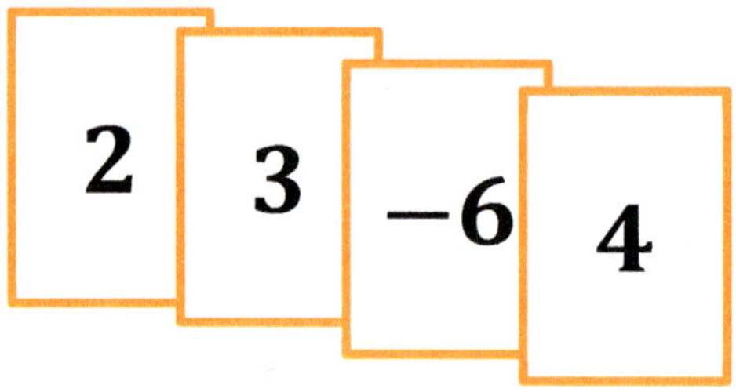

   a. If all of the matching cards will increase her score by 18, what are the matching cards?

   b. If all of the matching cards will decrease her score by 12, what are the matching cards?

2. A hand of six integer cards has one matching set of two or more cards. If the matching set of cards is removed from the hand, the score of the hand will increase by six. What are the possible values of these matching cards? Explain. Write an equation using multiplication showing how the matching cards yield an increase in score of six.

Exit Ticket Sample Solutions

1. **Natalie is playing the Integer Game and only shows you the four cards shown below. She tells you that the rest of her cards have the same values on them and match one of these four cards.**

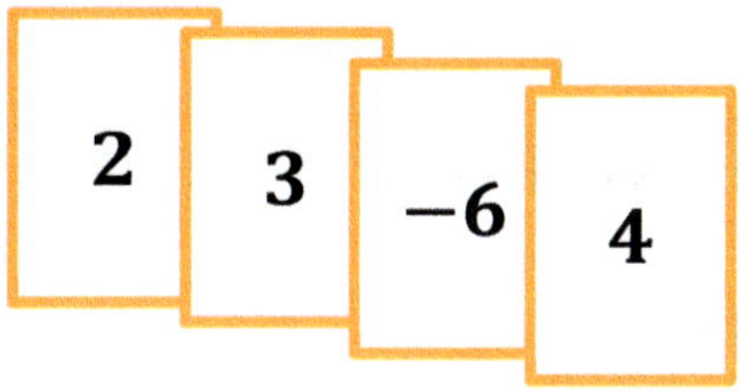

   a. **If all of the matching cards will increase her score by 18, what are the matching cards?**

   If there were nine 2 cards, then $2+2+2+2+2+2+2+2+2=18$

   $$9\times 2=18$$

   If there were six 3 cards, then $3+3+3+3+3+3=18$

   $$6\times 3=18$$

   b. **If all of the matching cards will decrease her score by 12, what are the matching cards?**

   If there were two -6 cards, then $(-6)+(-6)=-12$

   $$2\times(-6)=-12$$

2. **A hand of six integer cards has one matching set of two or more cards. If the matching set of cards is removed from the hand, the score of the hand will increase by six. What are the possible values of these matching cards? Explain. Write an equation using multiplication showing how the matching cards yield an increase in score of six.**

   If the matching cards are taken away from the playing hand and the score of the hand increases, then the matching cards must have negative values. The playing hand only has six cards, so the number of matching cards is limited to six. Taking away the following matching sets would increase the score by six:

   Taking away one set of two -3 cards can be represented by $-(-3)-(-3)$

   $$3+3=6$$

   $$3\times 2=6 \textit{ or } (-3)\times(-2)=6$$

   Taking away one set of three -2 cards can be represented by $-(-2)-(-2)-(-2)$

   $$2+2+2=6$$

   $$2\times 3=6 \textit{ or } (-2)\times(-3)=6$$

   Taking away one set of six -1 cards can be represented by $-(-1)-(-1)-(-1)-(-1)-(-1)-(-1)$

   $$1+1+1+1+1+1=6$$

   $$1\times 6=6 \textit{ or } (-1)\times(-6)=6$$

Problem Set Sample Solutions

1. Describe sets of two or more matching integer cards that satisfy the criteria in each part below:

   a. Cards increase the score by eight points.

      Picking up: eight 1*'s, four* 2*'s, or two* 4*'s*

      OR

      Removing: eight (-1)*'s, four* (-2)*'s, or two* (-4)*'s*

   b. Cards decrease the score by 9 points.

      Picking up: nine (-1)*'s or three* (-3)*'s*

      OR

      Removing: nine 1*'s or three* 3*'s*

   c. Removing cards that increase the score by 10 points.

      Ten (-1)*'s, five* (-2)*'s, or two* (-5)*'s*

   d. Positive cards that decrease the score by 18 points.

      Removing eighteen 1*'s, nine* 2*'s, six* 3*'s, three* 6*'s, or two* 9*'s.*

2. You have the integer cards shown at the right when your teacher tells you to choose a card to multiply four times. If your goal is to get your score as close to zero as possible, which card would you choose? Explain how your choice changes your score.

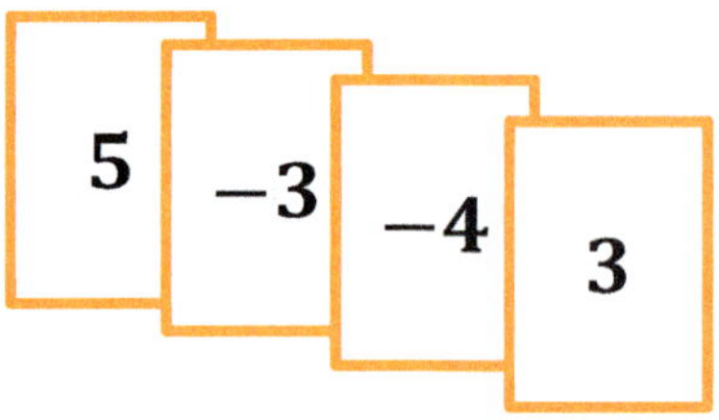

   The best choice to multiply is the -3*. The cards currently have a score of one. The new score with the* -3 *multiplied by* 4 *is* -8*. The scores where the other cards are multiplied by* 4 *are* 10*,* -11*, and* 16*, which are all further from zero.*

3. Sherry is playing the Integer Game and is given a chance to discard a set of matching cards. Sherry determines that if she discards one set of cards, her score will increase by 12. If she discards another set, then her score will decrease by eight. If her matching cards make up all six cards in her hand, what cards are in Sherry's hand? Are there any other possibilities?

   There are two possibilities:

   $2, 2, 2, 2, -6, -6$

   OR

   $-3, -3, -3, -3, 4, 4$

Lesson 11: Develop Rules for Multiplying Signed Numbers

Student Outcomes

- Students understand the rules for multiplication of integers and that multiplying the absolute values of integers results in the absolute value of the product. The sign, or absolute value, of the product is positive if the factors have the same sign and negative if they have opposite signs.
- Students realize that $(-1)(-1) = (1)$ and see that it can be proven mathematically using the distributive property and the additive inverse.
- Students use the rules for multiplication of signed numbers and give real-world examples.

Classwork

Example 1 (17 minutes): Extending Whole Number Multiplication to the Integers

Part A: Students complete only the right half of the table in the student materials. They do this by calculating the total change to a player's score using the various sets of matching cards. Students complete the table with these values to reveal patterns in multiplication.

Students describe, using Integer Game scenarios, what the right quadrants of the table represent and record this in the student materials.

Example 1: Extending Whole Number Multiplication to the Integers

Part A: Complete quadrants I and IV of the table below to show how sets of matching integer cards will affect a player's score in the Integer Game. For example, three 2's would increase a player's score by $0 + 2 + 2 + 2 = 6$ points.

Quadrant II | Quadrant I

-25	-20	-15	-10	-5	5	5	10	15	20	25
-20	-16	-12	-8	-4	4	4	8	12	16	20
-15	-12	-9	-6	-3	3	3	6	9	12	15
-10	-8	-6	-4	-2	2	2	4	6	8	10
-5	-4	-3	-2	-1	1	1	2	3	4	5
-5	-4	-3	-2	-1	0	1	2	3	4	5
5	4	3	2	1	-1	-1	-2	-3	-4	-5
10	8	6	4	2	-2	-2	-4	-6	-8	-10
15	12	9	6	3	-3	-3	-6	-9	-12	-15
20	16	12	8	4	-4	-4	-8	-12	-16	-20
25	20	15	10	5	-5	-5	-10	-15	-20	-25

Quadrant III | Quadrant IV

← Number of matching cards (middle row)

↑ Integer card values (middle column)

Quadrant II — What does this quadrant represent? *Removing positive value cards*

Quadrant I — What does this quadrant represent? *Picking up positive value cards*

Quadrant III — What does this quadrant represent? *Removing negative value cards*

Quadrant IV — What does this quadrant represent? *Picking up negative value cards*

a. **What patterns do you see in the right half of the table?**

The products in quadrant I are positive and the products in quadrant IV are negative. When looking at the absolute values of the products, quadrants I and IV are a reflection of each other with respect to the middle row.

b. **Enter the missing integers in the left side of the middle row, and describe what they represent.**

The numbers represent how many matching cards are being discarded or removed.

-5	-4	-3	-2	-1	0

Part B: Students complete quadrant II of the table.

Students describe, using an Integer Game scenario, what quadrant II of the table represents and record this in the student materials.

Part B: Complete quadrant II of the table.

Quadrant II

What does this quadrant represent?

Removing positive value cards.

-25	-20	-15	-10	-5	5
-20	-16	-12	-8	-4	4
-15	-12	-9	-6	-3	3
-10	-8	-6	-4	-2	2
-5	-4	-3	-2	-1	1
-5	-4	-3	-2	-1	0

Students answer the following questions:

MP.7

c. **What relationships or patterns do you notice between the products (values) in quadrant II and the products (values) in quadrant I?**

The products in quadrant II are all negative values. Looking at the absolute values of the products, quadrant I and II are a reflection of each other with respect to the center column.

d. **What relationships or patterns do you notice between the products (values) in quadrant II and the products (values) in quadrant IV?**

The products in quadrants II and IV are all negative values. Each product of integers in quadrant II is equal to the product of their opposites in quadrant IV.

e. **Use what you know about the products (values) in quadrants I, II, and IV to describe what quadrant III will look like when its products (values) are entered.**

The reflection symmetry of quadrant I to quadrants II and IV suggests that there should be similar relationships between quadrant II, III, and IV. The number patterns in quadrants II and IV also suggest that the products in quadrant III are positive values.

Part C: Discuss the following question. Then instruct students to complete the final quadrant of the table.

- In the Integer Game, what happens to a player's score when he removes a matching set of cards with negative values from his hand?
  - *His score increases because the negative cards that cause the score to decrease are removed.*

Students describe, using an Integer Game scenario, what quadrant III of the table represents and complete the quadrant in the student materials.

> *Scaffolding:*
> Create an *anchor poster* showing the quadrants with the new rules for multiplying integers.

Part C: Complete quadrant III of the table.

Refer to the completed table to help you answer the following questions:

What does this quadrant represent?

Removing negative value cards.

-5	-4	-3	-2	-1	0
5	4	3	2	1	-1
10	8	6	4	2	-2
15	12	9	6	3	-3
20	16	12	8	4	-4
25	20	15	10	5	-5

Quadrant III ↑

Students refer to the completed table to answer parts (f) and (g).

f. Is it possible to know the sign of a product of two integers just by knowing in which quadrant each integer is located? Explain.

Yes, it is possible to know the sign of a product of two integers just by knowing each integer's quadrant because the signs of the values in each of the quadrants are consistent. Two quadrants contain positive values, and the other two quadrants contain negative values.

g. Which quadrants contain which values? Describe an Integer Game scenario represented in each quadrant.

Quadrants I and III contain all positive values. Picking up three 4's is represented in quadrant I and increases your score. Removing three (-4)'s is represented in quadrant III and also increases your score. Quadrants II and IV contain all negative values. Picking up three (-4)'s is represented in quadrant IV and decreases your score. Removing three 4's is represented in quadrant II and also decreases your score.

Example 2 (10 minutes): Using Properties of Arithmetic to Explain Multiplication of Negative Numbers

The teacher guides students to verify their conjecture that the product of two negative integers is positive using the distributive property and the additive inverse property.

Example 2: Using Properties of Arithmetic to Explain Multiplication of Negative Numbers

- We have used the Integer Game to explain that adding a number multiple times has the same effect as removing the opposite value the same number of times. What is $(-1) \times (-1)$?
  - *Removing a -1 card is the same as adding a 1 card. So, $(-1) \times (-1) = 1$.*

> *Scaffolding:*
> Use color or highlight steps to help students organize and understand the manipulations.

- Why are 1 and -1 called additive inverses? Write an equation that shows this property.
  - *The sum of 1 and -1 is 0; $1 + (-1) = 0$.*

We are now going to show $-1 \times (-1) = 1$ using properties of arithmetic.

- We know $1 + (-1) = 0$ is true.
- We will show that $(-1) \times (-1)$ is the additive inverse of -1, which is 1.

If $-1 \times 0 = 0$	By the zero product property
then $-1 \times \big(1 + (-1)\big) = 0$	By substitution of $\big(1 + (-1)\big)$ for 0
$(-1 \times 1) + \big(-1 \times (-1)\big) = 0$	Distributive property
$-1 + \big(-1 \times (-1)\big) = 0$	Multiplication by 1

- Since -1 and $\big(-1 \times (-1)\big)$ have a sum of zero, they are additive inverses of each other; but, the additive inverse of -1 is 1.
- Because $\big(-1 \times (-1)\big)$ is the additive inverse of -1, we conclude that $(-1) \times (-1) = 1$. This fact can be used to show that $-1 \times a = -a$ for any integer and that $-a \times b = -(a \times b)$ for any integers a and b.

Exercise 1 (8 minutes): Multiplication of Integers in the Real World

Students create real-world scenarios for expressions given in the student materials. Students may use an Integer Game scenario as a reference.

Exercise 1: Multiplication of Integers in the Real World

Generate real-world situations that can be modeled by each of the following multiplication problems. Use the Integer Game as a resource.

a. -3×5

I lost three $\$5$ bills, and now I have $-\$15$.

b. $-6 \times (-3)$

I removed six (-3)'s from my hand in the Integer Game, and my score increased 18 points.

c. $4 \times (-7)$

If I lose 7 lb. per month for 4 months, my weight will change -28 lb. total.

> *Scaffolding:*
> For English language learners, create a teacher/student T-chart on which the teacher writes a real-world situation that corresponds with a product, and students write similar situations using different numbers.

Closing (5 minutes)

- How do we determine if the product of two signed numbers will be positive or negative?
  - *If the factors have the same sign, the product will be positive, and if the factors have opposite signs, the product will be negative.*
- Why does the product of two negative values result in a positive value? Explain using the Integer Game.
  - *The product of two negative numbers represents removing negative cards during the Integer Game. When negative cards are removed from someone's hand, the score increases; therefore, the product is positive.*

> Lesson Summary
>
> **To multiply signed numbers, multiply the absolute values to get the absolute value of the product. The sign of the product is positive if the factors have the same sign and negative if they have opposite signs.**

Exit Ticket (5 minutes)

Name ____ Date ________________

Lesson 11: Develop Rules for Multiplying Signed Numbers

Exit Ticket

1. Create a real-life example that can be modeled by the expression -2×4, and then state the product.

2. Two integers are multiplied, and their product is a positive number. What must be true about the two integers?

Exit Ticket Sample Solutions

1. **Create a real-life example that can be modeled by the expression -2×4, and then state the product.**

   Answers will vary. Tobi wants to lose 2 lb. each week for four weeks. Write an integer to represent Tobi's weight change after four weeks. Tobi's weight changes by -8 lb. after four weeks.

2. **Two integers are multiplied, and their product is a positive number. What must be true about the two integers?**

   Both integers must be positive numbers, or both integers must be negative numbers.

Problem Set Sample Solutions

1. **Complete the problems below. Then, answer the question that follows.**

$3 \times 3 = 9$	$3 \times 2 = 6$	$3 \times 1 = 3$	$3 \times 0 = 0$	$3 \times (-1) = -3$	$3 \times (-2) = -6$
$2 \times 3 = 6$	$2 \times 2 = 4$	$2 \times 1 = 2$	$2 \times 0 = 0$	$2 \times (-1) = -2$	$2 \times (-2) = -4$
$1 \times 3 = 3$	$1 \times 2 = 2$	$1 \times 1 = 1$	$1 \times 0 = 0$	$1 \times (-1) = -1$	$1 \times (-2) = -2$
$0 \times 3 = 0$	$0 \times 2 = 0$	$0 \times 1 = 0$	$0 \times 0 = 0$	$0 \times (-1) = 0$	$0 \times (-2) = 0$
$-1 \times 3 = -3$	$-1 \times 2 = -2$	$-1 \times 1 = -1$	$-1 \times 0 = 0$	$-1 \times (-1) = 1$	$-1 \times (-2) = 2$
$-2 \times 3 = -6$	$-2 \times 2 = -4$	$-2 \times 1 = -2$	$-2 \times 0 = 0$	$-2 \times (-1) = 2$	$-2 \times (-2) = 4$
$-3 \times 3 = -9$	$-3 \times 2 = -6$	$-3 \times 1 = -3$	$-3 \times 0 = 0$	$-3 \times (-1) = 3$	$-3 \times (-2) = 6$

   Which row shows the same pattern as the outlined column? Are the problems similar or different? Explain.

   The row outlined shows the same pattern as the outlined column. The problems have the same answers, but the signs of the integer factors are switched. For example, $3 \times (-1) = -3 \times 1$. This shows that the product of two integers with opposite signs is equal to the product of their opposites.

2. **Explain why $(-4) \times (-5) = 20$. Use patterns, an example from the Integer Game, or the properties of operations to support your reasoning.**

   Answers may vary. Losing four -5 cards will increase a score in the Integer Game by 20 because a negative value decreases a score, but the score increases when it is removed.

3. **Each time that Samantha rides the commuter train, she spends \$4 for her fare. Write an integer that represents the change in Samantha's money from riding the commuter train to and from work for 13 days. Explain your reasoning.**

   If Samantha rides to and from work for 13 days, then she rides the train a total of 26 times. The cost of each ride can be represented by -4. So, the change to Samantha's money is represented by $-4 \times 26 = -104$. The change to Samantha's money after 13 days of riding the train to and from work is $-\$104$.

4. **Write a real-world problem that can be modeled by $4 \times (-7)$.**

   Answers will vary. Every day, Alec loses 7 pounds of air pressure in a tire on his car. At that rate, what is the change in air pressure in his tire after 4 days?

Challenge:

5. Use properties to explain why for each integer a, $-a = -1 \times a$. (Hint: What does $(1 + (-1)) \times a$ equal? What is the additive inverse of a?)

   $0 \times a = 0$ *Zero product property*

   $(1 + (-1)) \times a = 0$ *Substitution*

   $a + (-1 \times a) = 0$ *Distributive property*

   Since a and $(-1 \times a)$ have a sum of zero, they must be additive inverses. By definition, the additive inverse of a is $-a$, so $(-1 \times a) = -a$.

Lesson 12: Division of Integers

Student Outcomes

- Students recognize that division is the reverse process of multiplication and that integers can be divided provided the divisor is not zero. If p and q are integers, then $-\left(\frac{p}{q}\right) = \frac{-p}{q} = \frac{p}{-q}$.
- Students understand that every quotient of integers (with a nonzero divisor) is a rational number and divide signed numbers by dividing their absolute values to get the absolute value of the quotient. The quotient is positive if the divisor and dividend have the same signs and negative if they have opposite signs.

Classwork

Exercise 1 (5 minutes): Recalling the Relationship Between Multiplication and Division

MP.2

The teacher gives each student a card with a whole number multiplication or division math fact on it. Students move around the room in search of other students who have related math facts. (If the class size does not allow for exact multiples of 4, then extra cards may be placed on desk tops for students to find.) Four cards make a *match* (e.g., $6 \times 4 = 24$, $4 \times 6 = 24$, $24 \div 6 = 4$, and $24 \div 4 = 6$). After four students locate each other, they sit down together and record the equations from their cards into their student materials as indicated below. The teacher circulates among students as a facilitator, guiding those who are having trouble. Once all groups are formed and each group has shared its related facts with the class, the teacher collects the fact cards and directs students back to their original seats.

Exercise 1: Recalling the Relationship Between Multiplication and Division

Record equations from Exercise 1 on the left.

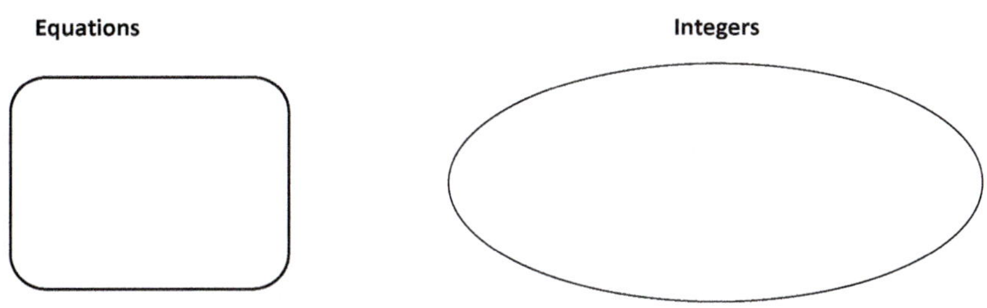

Example 1 (15 minutes): Transitioning from Integer Multiplication Rules to Integer Division Rules

Students make an *integer multiplication facts bubble* by expanding upon the four related math facts they wrote down.

Step 1: Students construct three similar integer multiplication problems, two problems using one negative number as a factor and one with both negative numbers as factors. Students may use the commutative property to extend their three equations to six.

> *Scaffolding:*
> Provide an example of a completed integer bubble for students who are struggling with the task.

Example 1: Transitioning from Integer Multiplication Rules to Integer Division Rules

Record your group's number sentences in the space on the left below.

Equations

$4 \times 6 = 24$

$6 \times 4 = 24$

$24 \div 4 = 6$

$24 \div 6 = 4$

Integers

$-6 \times 4 = -24$ *or* $4 \times -6 = -24$

$-4 \times 6 = -24$ *or* $6 \times -4 = -24$

$-4 \times (-6) = 24$ *or* $-6 \times -4 = 24$

Step 2: Students use the integer multiplication facts in their integer bubble to create six related integer division facts. Group members should discuss the inverse relationship and the resulting division fact that must be true based on each multiplication equation.

Equations

$4 \times 6 = 24$

$6 \times 4 = 24$

$24 \div 4 = 6$

$24 \div 6 = 4$

Integers

$-6 \times 4 = -24 \rightarrow -24 \div (-6) = 4$

$-24 \div 4 = -6$

$-4 \times 6 = -24 \rightarrow -24 \div (-4) = 6$

$-24 \div 6 = -4$

$-4 \times (-6) = 24 \rightarrow 24 \div (-4) = -6$

$24 \div (-6) = -4$

MP.8 Step 3: Students use the equations in their integer bubble and the patterns they observed to answer the following questions.

a. **List examples of division problems that produced a quotient that is a negative number.**

$-24 \div 4 = -6$; $-24 \div 6 = -4$; $24 \div (-4) = -6$; $24 \div (-6) = -4$

b. **If the quotient is a negative number, what must be true about the signs of the dividend and divisor?**

The quotient is a negative number when the signs of the dividend and divisor are not the same; one is positive, and one is negative.

c. **List your examples of division problems that produced a quotient that is a positive number.**

$-24 \div (-4) = 6$; $-24 \div (-6) = 4$; $24 \div 4 = 6$; $24 \div 6 = 4$

d. **If the quotient is a positive number, what must be true about the signs of the dividend and divisor?**

The quotient is a positive number when the signs of the dividend and the divisor are the same in each case.

Step 4: Whole-group discussion. Students share answers from Step 3 with the class. The class comes to a consensus and realizes that since multiplication and division are related* (inverse operations), the integer rules for these operations are related. Students summarize the rules for division, which are stated in the Lesson Summary of the student materials. (*Reminder: The rules apply to all situations except dividing by zero.)

Rules for Dividing Two Integers:

- **A quotient is negative if the divisor and the dividend have** ___*opposite*___ **signs.**
- **A quotient is positive if the divisor and the dividend have** ___*the same*___ **signs.**

Exercise 2 (8 minutes): Is the Quotient of Two Integers Always an Integer?

MP.3

Students explore the question above by coming up with an example to prove or refute their position.

Allow 3–5 minutes for students to create a math example or *counterexample*, along with a written response to support their position. Students present their cases to the class.

Exercise 2: Is the Quotient of Two Integers Always an Integer?

Is the quotient of two integers always an integer? Use the work space below to create quotients of integers. Answer the question, and use examples or a counterexample to support your claim.

Work Space:

$-24 \div 6 = -4$ — *Example of an integer quotient*

$6 \div (-24) = \frac{6}{-24} = \frac{1}{-4} = -\frac{1}{4}$ — *Counterexample: has a non-integer quotient*

Answer:

No, quotients of integers are not always integers. In my first example above, $-24 \div 6$ yields an integer quotient -4. However, when I switched the divisor and dividend, that quotient divides a number with a smaller absolute value by a number with a greater absolute value, making the quotient a rational number between -1 and 1. In dividing $6 \div (-24)$, the quotient is $\frac{6}{-24} = \frac{1}{-4}$. Of course, $-\frac{1}{4}$ is not an integer but is the opposite value of the fraction $\frac{1}{4}$. This counterexample shows that quotients of integers are not always integers.

Conclusion: Every quotient of two integers is always a rational number but not always an integer.

Once students have disproved the statement with a counterexample (where the quotient is a decimal or fraction), ask students to determine what must be true of two integers if their quotient *is* an integer. Students may need some time to study the examples where the quotient is an integer to determine that the quotient of two integers, $\frac{A}{B}$, $B \neq 0$, is an integer when either $B = 1$ or $A = kB$ for any integer k.

Exercise 3 (5 minutes): Different Representations of the Same Quotient

Students are given the three different representations below and must determine the answers. Are the answers the same or different? Why or why not? Allow time for students to answer with their groups or learning partner before addressing this in the form of a whole-group discussion.

Exercise 3: Different Representation of the Same Quotient

Are the answers to the three quotients below the same or different? Why or why not?

a. $-14 \div 7$

$-14 \div 7 = -2$

b. $14 \div (-7)$

$14 \div (-7) = -2$

c. $-(14 \div 7)$

$-(14 \div 7) = -(2) = -2$

The answers to the problems are the same: -2. In part (c), the negative in front of the parentheses changes the value inside the parentheses to its opposite. The value in the parentheses is 2, and the opposite of 2 is -2.

Scaffolding:

For part (c), remind students to think about how a negative sign in front of a number or expression means the opposite.

Fluency Exercise (2 minutes): Integer Division

MP.7 See attached hand-out. Students answer as many questions as possible in one minute. One minute is allocated for going over the answers and recognizing achievements.

Fluency Exercise: Integer Division

1.	$-56 \div (-7) = 8$	**15.**	$-28 \div (-7) = 4$	**29.**	$-14 \div (-7) = 2$
2.	$-56 \div (-8) = 7$	**16.**	$-28 \div (-4) = 7$	**30.**	$-14 \div (-2) = 7$
3.	$56 \div (-8) = -7$	**17.**	$28 \div 4 = 7$	**31.**	$14 \div (-2) = -7$
4.	$-56 \div 7 = -8$	**18.**	$-28 \div 7 = -4$	**32.**	$-14 \div 7 = -2$
5.	$-40 \div (-5) = 8$	**19.**	$-20 \div (-5) = 4$	**33.**	$-10 \div (-5) = 2$
6.	$-40 \div (-4) = 10$	**20.**	$-20 \div (-4) = 5$	**34.**	$-10 \div (-2) = 5$
7.	$40 \div (-4) = -10$	**21.**	$20 \div (-4) = -5$	**35.**	$10 \div (-2) = -5$
8.	$-40 \div 5 = -8$	**22.**	$-20 \div 5 = -4$	**36.**	$-10 \div 5 = -2$
9.	$-16 \div (-4) = 4$	**23.**	$-8 \div (-4) = 2$	**37.**	$-4 \div (-4) = 1$
10.	$-16 \div (-2) = 8$	**24.**	$-8 \div (-2) = 4$	**38.**	$-4 \div (-1) = 4$
11.	$16 \div (-2) = -8$	**25.**	$8 \div (-2) = -4$	**39.**	$4 \div (-1) = -4$
12.	$-16 \div 4 = -4$	**26.**	$-8 \div 4 = -2$	**40.**	$-4 \div 1 = -4$
13.	$-3 \div (-4) = 0.75$	**27.**	$4 \div (-8) = -0.5$	**41.**	$1 \div (-4) = -0.25$
14.	$-3 \div 4 = -0.75$	**28.**	$-4 \div 8 = -0.5$	**42.**	$-1 \div 4 = -0.25$

Closing (5 minutes)

- How are the rules for multiplying integers and dividing integers related?
  - *The rules for multiplying integers and dividing integers are the same, as long as the divisor is not zero.*
- If I have a negative quotient, what must be true about the signs of the dividend and/or divisor?
  - *If the quotient is negative, the dividend and divisor must have opposite signs.*
- If I have a positive quotient, what must be true about the signs of the dividend and/or divisor?
  - *If the quotient is positive, the dividend and divisor must have the same sign.*

> Lesson Summary
>
> **The rules for dividing integers are similar to the rules for multiplying integers (when the divisor is not zero). The quotient is positive if the divisor and dividend have the same signs and negative if they have opposite signs.**
>
> **The quotient of any 2 integers (with a nonzero divisor) will be a rational number. If p and q are integers, then $-\left(\frac{p}{q}\right) = \frac{-p}{q} = \frac{p}{-q}$.**

MP.3

Exit Ticket (5 minutes)

Students determine whether or not various representations of the quotient of two integers are equivalent.

Name ______________________________ Date ______________

Lesson 12: Division of Integers

Exit Ticket

1. Mrs. McIntire, a seventh-grade math teacher, is grading papers. Three students gave the following responses to the same math problem:

   Student one: $\frac{1}{-2}$

   Student two: $-\left(\frac{1}{2}\right)$

   Student three: $-\frac{1}{2}$

   On Mrs. McIntire's answer key for the assignment, the correct answer is -0.5. Which student answer(s) is (are) correct? Explain.

2. Complete the table below. Provide an answer for each integer division problem, and write a related equation using integer multiplication.

Integer Division Problem	Related Equation Using Integer Multiplication
$-36 \div (-9) =$ ______	
$24 \div (-8) =$ ______	
$50 \div 10 =$ ______	
$42 \div 6 =$ ______	

EUREKA MATH™

Exit Ticket Sample Solutions

1. Mrs. McIntire, a seventh-grade math teacher, is grading papers. Three students gave the following responses to the same math problem:

   Student one: $\frac{1}{-2}$

   Student two: $-\left(\frac{1}{2}\right)$

   Student three: $-\frac{1}{2}$

   On Mrs. McIntire's answer key for the assignment, the correct answer is -0.5. Which student answer(s) is (are) correct? Explain.

   All student answers are correct since they are all equivalent to -0.5.

   For student one: $\frac{1}{-2}$ *means* 1 *divided by* -2. *When dividing a positive* 1 *by a negative* 2, *the answer will be negative five-tenths or* -0.5.

   For student two: $-\left(\frac{1}{2}\right)$ *means the opposite of* $\frac{1}{2}$. *One-half is equivalent to five-tenths, and the opposite is negative five-tenths or* -0.5.

   For student three: $-\frac{1}{2}$ *means* -1 *divided by* 2. *When dividing a negative* 1 *by a positive* 2, *the answer will be negative five-tenths or* -0.5.

2. Complete the table below. Provide an answer for each integer division problem, and write a related equation using integer multiplication.

Integer Division Problem	Related Equation Using Integer Multiplication
$-36 \div (-9) =$ <u>4</u>	$-9 \times 4 = -36$ *or* $4 \times (-9) = -36$
$24 \div (-8) =$ <u>-3</u>	$-8 \times (-3) = 24$ *or* $-3 \times (-8) = 24$
$-50 \div 10 =$ <u>-5</u>	$-5 \times 10 = -50$ *or* $10 \times (-5) = -50$
$42 \div 6 =$ <u>7</u>	$6 \times 7 = 42$ *or* $7 \times 6 = 42$

Problem Set Sample Solutions

1. **Find the missing values in each column.**

Column A	Column B	Column C	Column D
$48 \div 4 = 12$	$24 \div 4 = 6$	$63 \div 7 = 9$	$21 \div 7 = 3$
$-48 \div (-4) = 12$	$-24 \div (-4) = 6$	$-63 \div (-7) = 9$	$-21 \div (-7) = 3$
$-48 \div 4 = -12$	$-24 \div 4 = -6$	$-63 \div 7 = -9$	$-21 \div 7 = -3$
$48 \div (-4) = -12$	$24 \div (-4) = -6$	$63 \div (-7) = -9$	$21 \div (-7) = -3$

2. **Describe the pattern you see in each column's answers in Problem 1, relating it to the problems' divisors and dividends. Why is this so?**

The pattern in the columns' answers is the same two positive values followed by the same two negative values. This is so for the first two problems because the divisor and the dividend have the same signs and absolute values, which yields a positive quotient. This is so for the second two problems because the divisor and dividend have different signs but the same absolute values, which yields a negative quotient.

3. **Describe the pattern you see between the answers for Columns A and B in Problem 1 (e.g., compare the first answer in Column A to the first answer in Column B; compare the second answer in Column A to the second answer in Column B). Why is this so?**

The answers in Column B are each one-half of the corresponding answers in Column A. That is because the dividend of 48 in Column A is divided by 4, and the dividend of 24 in Column B is divided by 4 (and so on with the same order and same absolute values but different signs). Since 24 is half of 48, the quotient (answer) in Column B will be one-half of the quotient in Column A.

4. **Describe the pattern you see between the answers for Columns C and D in Problem 1. Why is this so?**

The answers in Column D are each one-third of the corresponding answers in Column C. That is because the dividend of 63 in Column C is divided by 7, and the dividend of 21 in Column D is divided by 7 (and so on with the same order and same absolute values but different signs). Since 21 is one-third of 63, the quotient (answer) in Column D will be one-third of the quotient in Column C.

Fluency Exercise: Integer Division

1.	$-56 \div (-7) =$	15.	$-28 \div (-7) =$	29.	$-14 \div (-7) =$
2.	$-56 \div (-8) =$	16.	$-28 \div (-4) =$	30.	$-14 \div (-2) =$
3.	$56 \div (-8) =$	17.	$28 \div 4 =$	31.	$14 \div (-2) =$
4.	$-56 \div 7 =$	18.	$-28 \div 7 =$	32.	$-14 \div 7 =$
5.	$-40 \div (-5) =$	19.	$-20 \div (-5) =$	33.	$-10 \div (-5) =$
6.	$-40 \div (-4) =$	20.	$-20 \div (-4) =$	34.	$-10 \div (-2) =$
7.	$40 \div (-4) =$	21.	$20 \div (-4) =$	35.	$10 \div (-2) =$
8.	$-40 \div 5 =$	22.	$-20 \div 5 =$	36.	$-10 \div 5 =$
9.	$-16 \div (-4) =$	23.	$-8 \div (-4) =$	37.	$-4 \div (-4) =$
10.	$-16 \div (-2) =$	24.	$-8 \div (-2) =$	38.	$-4 \div (-1) =$
11.	$16 \div (-2) =$	25.	$8 \div (-2) =$	39.	$4 \div (-1) =$
12.	$-16 \div 4 =$	26.	$-8 \div 4 =$	40.	$-4 \div 1 =$
13.	$-3 \div (-4) =$	27.	$4 \div (-8) =$	41.	$1 \div (-4) =$
14.	$-3 \div 4 =$	28.	$-4 \div 8 =$	42.	$-1 \div 4 =$

Lesson 13: Converting Between Fractions and Decimals Using Equivalent Fractions

Student Outcomes

- Students understand that the context of a real-life situation often determines whether a rational number should be represented as a fraction or decimal.
- Students understand that decimals specify points on the number line by repeatedly subdividing intervals into tenths (*deci-* means tenth).
- Students convert positive decimals to fractions and fractions to decimals when the denominator is a product of only factors of either 2 or 5.

Classwork

Opening Exercise (4 minutes)

As was seen in Lesson 12, when dividing many integers, the result is a non-integer quotient. These types of numbers are evident in the real world. For an Opening Exercise, direct students as they enter the room to provide responses to each of two questions posted on poster paper (questions listed below) using sticky notes.

The two questions to post are as follows:

1. What are some examples from the real world where decimals are used?

   Possible answers: Money, metric system, etc.

2. What are some examples from the real world where fractions are used?

   Possible answers: Some measurement (carpentry, cooking, etc.)

Discuss appropriate responses as a class; then ask the following questions aloud:

- Have you ever seen a recipe call for 2.7 cups of flour? Why or why not?
  - *Measuring cups for cooking are generally labeled with $\frac{1}{2}$, $\frac{1}{3}$, $\frac{1}{4}$, etc., for recipes requiring measurements in fractional cups.*
- How do you think people would react if a local gas station posted the price of gasoline as $3\frac{3}{7}$ dollars per gallon? Why?
  - *Dollars are never measured as $\frac{1}{6}$ or $\frac{1}{3}$ of a dollar; dollars are measured in decimal form using tenths and hundredths.*

Example 1 (1 minute): Representations of Rational Numbers in the Real World

Students describe in their own words why they need to know how to represent rational numbers in different ways.

Example 1: Representations of Rational Numbers in the Real World

Following the Opening Exercise and class discussion, describe why we need to know how to represent rational numbers in different ways.

Different situations in the real world require different representations of rational numbers. Because of common usage in life outside of the classroom, we may automatically know that a quarter of a dollar is the same as 25 *cents, or a quarter, but for people who are used to measuring money in only decimals, a quarter of a dollar might not make much sense.*

Example 2 (10 minutes): Using Place Values to Write (Terminating) Decimals as Equivalent Fractions

MP.2

Students use the place value of the right-most decimal place in a terminating decimal to rewrite a positive rational number as an equivalent fraction.

Example 2: Using Place Values to Write (Terminating) Decimals as Equivalent Fractions

a. **What is the value of the number** 2.25**? How can this number be written as a fraction or mixed number?**

Two and twenty-five hundredths or $2\frac{25}{100}$

Scaffolding:

Provide or create a place value chart to aid those who do not remember their place values or for English language learners who are unfamiliar with the vocabulary.

- How do we rewrite this fraction (or any fraction) in its simplest form?
  - *If a factor is common to both the numerator and denominator of a fraction, the fraction can be simplified, resulting in a fraction whose numerator and denominator only have a common factor of* 1 *(the numerator and denominator are relatively prime).*

b. **Rewrite the fraction in its simplest form showing all steps that you use.**

$\frac{25}{100} = \frac{25}{4 \times 25} = \frac{1}{4} \quad \rightarrow \quad 2\frac{25}{100} = 2\frac{1}{4}$

c. **What is the value of the number** 2.025**? How can this number be written as a mixed number?**

Two and twenty-five thousandths, or $2\frac{25}{1{,}000}$

d. **Rewrite the fraction in its simplest form showing all steps that you use.**

$\frac{25}{1{,}000} = \frac{25}{100 \times 10} \qquad \frac{25}{4 \times 25 \times 10} = \frac{1}{40} \quad \rightarrow \quad 2\frac{25}{1{,}000} = 2\frac{1}{40}$

Scaffolding:

Have students create a graphic organizer to relate the different representations of rational numbers , including fractions, decimals, and words. Pictures may also be used if applicable.

Exercise 1 (5 minutes)

Exercise 1

Use place value to convert each terminating decimal to a fraction. Then rewrite each fraction in its simplest form.

a. 0.218

$\frac{218}{1{,}000} = \frac{109\times2}{500\times2} = \frac{109}{500} \rightarrow 0.218 = \frac{109}{500}$

b. 0.16

$\frac{16}{100} = \frac{4\times4}{4\times25} = \frac{4}{25} \rightarrow 0.16 = \frac{4}{25}$

c. 2.72

$\frac{72}{100} = \frac{4\times18}{4\times25} = \frac{18}{25} \rightarrow 2.72 = 2\frac{18}{25}$

d. 0.0005

$\frac{5}{10{,}000} = \frac{5\times1}{5\times2{,}000} = \frac{1}{2{,}000} \rightarrow 0.0005 = \frac{1}{2{,}000}$

- What do you notice about the denominators of fractions that represent each decimal place?
  - *The denominators are all powers of* 10.
- What are the prime factors of 10? 100? 1,000?
  - $10 = 2 \times 5$ $\quad 100 = 2^2 \times 5^2$ $\quad 1{,}000 = 2^3 \times 5^3$
    $10^1 = 2 \times 5$ $\quad 10^2 = 2^2 \times 5^2$ $\quad 10^3 = 2^3 \times 5^3$
- What prime factors make up the powers of 10?
  - *The powers of* 10 *contain only the factors* 2 *and* 5*, and, in each case, the number of factors of* 2 *and* 5 *are equal to the number of factors of* 10.
- How can the prime factorization of the powers of ten be used to write fractions in decimal form?
  - *Find an equivalent fraction whose denominator is a power of ten, then write the decimal representation using place values.*

Example 3 (10 minutes): Converting Fractions to Decimals—Fractions with Denominators Having Factors of only 2 or 5

Discuss the meaning of the term *decimal* as it is derived from the Latin word *decimus*, meaning *tenth*.

- What is the meaning of *tenth?* Provide real-world examples where *tenths* are regularly used.
  - *If a unit has been divided into ten equal-sized pieces, then one-tenth is the value of one of those ten pieces. A dime is one-tenth of a dollar; a penny is one-tenth of a dime.*

> *Scaffolding:*
>
> The prefix *deci-* is used in the metric system of measurement in which its meaning is one-tenth of a unit.

Students use equivalent fractions whose denominators include only the factors 2 and 5 to write decimal representations of rational numbers.

Example 3: Converting Fractions to Decimals—Fractions with Denominators Having Factors of only 2 or 5

a. **What are *decimals*?**

Decimals specify points on the number line by repeatedly subdividing intervals into tenths. If a unit is divided into ten equal-sized pieces, one piece would be one-tenth of that unit.

b. **Use the meaning of *decimal* to relate decimal place values.**

Each place value in a decimal is $\frac{1}{10}$ *of the value of the place to its left. This means that the denominators of the fractions that represent each decimal place value must be powers of ten.*

c. **Write the number $\frac{3}{100}$ as a decimal. Describe your process.**

The decimal form is 0.03*. The fraction includes a power of ten,* 100*, as its denominator. The value of the second decimal place is* $\frac{1}{100}$*, so* $\frac{3}{100}$ *in decimal form is* 0.03*.*

- How could we obtain an equivalent fraction to $\frac{3}{20}$ with a power of ten in the denominator?
    - *If there was another factor of* 5 *in the denominator, then we would have an equal number of* 2*'s and* 5*'s resulting in a power of ten. If we multiply the fraction by* $\frac{5}{5}$ *(or* 1*), we get an equivalent fraction with a power of ten in its denominator.*

d. **Write the number $\frac{3}{20}$ as a decimal. Describe your process.**

The fractional form is $\frac{3}{20} = \frac{3}{2^2\times 5}$*. The denominator lacks a factor of* 5 *to be a power of ten. To arrive at the decimal form, I multiply the fractional form by* $\frac{5}{5}$ *to arrive at* $\frac{3}{2^2\times 5} \times \frac{5}{5} = \frac{3\times 5}{2^2\times 5^2} = \frac{15}{100}$*, and* $\frac{15}{100} = 0.15$*.*

e. **Write the number $\frac{10}{25}$ as a decimal. Describe your process.**

The fractional form is $\frac{10}{25} = \frac{2\times 5}{5\times 5}$*; and, since* $\frac{5}{5} = 1$*, then* $\frac{2\times 5}{5\times 5} = \frac{2}{5}$*. The denominator lacks a factor of* 2 *to be a power of ten. To arrive at the decimal form, I multiply the fractional form by* $\frac{2}{2}$ *to arrive at* $\frac{2}{5} \times \frac{2}{2} = \frac{4}{10}$*, and* $\frac{4}{10} = 0.4$*.*

f. **Write the number $\frac{8}{40}$ as a decimal. Describe your process.**

The fractional form is $\frac{8}{40} = \frac{2^3}{2^3\times 5}$*. There are factors of* 2 *in the numerator and denominator that will cancel. If I leave one factor of two in the denominator, it will be* 10 *(a power of ten).*

$$\frac{2^3}{2^3 \times 5} = \frac{2^2 \times 2}{2^2 \times 2 \times 5} = \frac{2}{2 \times 5} = \frac{2}{10} = 0.2$$

Exercise 2 (5 minutes)

Students convert fractions to decimal form using equivalent fractions.

Exercise 2

Convert each fraction to a decimal using an equivalent fraction.

a. $\frac{3}{16} =$

$\frac{3}{16} = \frac{3}{2^4} \rightarrow \frac{3 \times 5^4}{2^4 \times 5^4} = \frac{1,875}{10,000} \rightarrow$

$\frac{1875}{10,000} = 0.1875$

b. $\frac{7}{5} =$

$\frac{7}{5} \rightarrow \frac{7 \times 2}{5 \times 2} = \frac{14}{10} \rightarrow \frac{14}{10} = 1\frac{4}{10} = 1.4$

c. $\frac{11}{32} =$

$\frac{11}{32} = \frac{11}{2^5} \rightarrow \frac{11 \times 5^5}{2^5 \times 5^5} = \frac{34,375}{100,000} \rightarrow$

$\frac{34375}{100,000} = 0.34375$

d. $\frac{35}{50} =$

$\frac{35}{50} = \frac{5 \times 7}{5^2 \times 2} \rightarrow \frac{7}{5 \times 2} = \frac{7}{10} \rightarrow \frac{7}{10} = 0.7$

Closing (5 minutes)

The closing questions reinforce the important ideas in the lesson.

- When asked to write a decimal value as a fraction (or mixed number), how do we determine the value of the denominator?
  - *The place value of the right-most decimal place shares the same denominator as an equivalent fraction representing the decimal.*
- If the denominator of a fraction in its simplest form has four factors of 2 and seven factors of 5, describe two different ways in which a power of ten can be obtained in the denominator.
  - *Three factors of* 2 *could be multiplied in to obtain an equivalent fraction, or three factors of* 5 *could be divided out to obtain a different equivalent fraction.*
- Consider for Lesson 14: Do you think it is possible to write a fraction whose denominator has factors other than 2 and 5 as a decimal?

Lesson Summary

Any terminating decimal can be converted to a fraction using place value (e.g., 0.35 is thirty-five hundredths or $\frac{35}{100}$). A fraction whose denominator includes only factors of 2 and 5 can be converted to a decimal by writing the denominator as a power of ten.

Exit Ticket (5 minutes)

Name ______________________________ Date ______________

Lesson 13: Converting Between Fractions and Decimals Using Equivalent Fractions

Exit Ticket

1. Write 3.0035 as a fraction. Explain your process.

2. This week is just one of 40 weeks that you spend in the classroom this school year. Convert the fraction $\frac{1}{40}$ to decimal form.

Exit Ticket Sample Solutions

1. Write 3.0035 as a fraction. Explain your process.

   The right-most decimal place is the ten-thousandths place, so the number in fractional form would be $3\frac{35}{10{,}000}$*. There are common factors of* 5 *in the numerator and denominator, and dividing both by these results in the fraction* $3\frac{7}{2{,}000}$.

2. This week is just one of 40 weeks that you spend in the classroom this school year. Convert the fraction $\frac{1}{40}$ to decimal form.

   $$\frac{1}{40}=\frac{1}{2^3\times 5}\times\frac{5^2}{5^2}=\frac{5^2}{2^3\times 5^3}=\frac{25}{1{,}000}=0.025$$

Scaffolding:

Extend Exit Ticket Problem 2 by asking students to represent this week as a percentage of the school year.

Answer: 2.5%

Problem Set Sample Solutions

1. Convert each terminating decimal to a fraction in its simplest form.

   a. 0.4

   $0.4=\frac{2}{5}$

   b. 0.16

   $0.16=\frac{4}{25}$

   c. 0.625

   $0.625=\frac{5}{8}$

   d. 0.08

   $0.08=\frac{2}{25}$

   e. 0.012

   $0.012=\frac{3}{250}$

2. Convert each fraction or mixed number to a decimal using an equivalent fraction.

   a. $\frac{4}{5}$

   $\frac{4}{5}=0.8$

   b. $\frac{3}{40}$

   $\frac{3}{40}=0.075$

   c. $\frac{8}{200}$

   $\frac{8}{200}=0.04$

   d. $3\frac{5}{16}$

   $3\frac{5}{16}=3.3125$

3. Tanja is converting a fraction into a decimal by finding an equivalent fraction that has a power of 10 in the denominator. Sara looks at the last step in Tanja's work (shown below) and says that she cannot go any further. Is Sara correct? If she is, explain why. If Sara is incorrect, complete the remaining steps.

   $$\frac{72}{480}=\frac{2^3\cdot 3^2}{2^5\cdot 3\cdot 5}$$

   Tanja can finish the conversion since there is a factor pair of 3*'s in the numerator and denominator that can be divided out with a quotient of* 1.

   Remaining Steps:

   $$\frac{720}{480}=\frac{2^3\cdot 3^2}{2^5\cdot 3\cdot 5}=\frac{3}{2^2\cdot 5}\left(\frac{5}{5}\right)=\frac{3\cdot 5}{2^2\cdot 5^2}=\frac{15}{100}$$

   Answer: 0.15

Lesson 14: Converting Rational Numbers to Decimals Using Long Division

Student Outcomes

- Students understand that every rational number can be converted to a decimal.
- Students represent fractions as decimal numbers that either terminate in zeros or repeat. Students then also represent repeating decimals using a bar over the shortest sequence of repeating digits.
- Students interpret word problems and convert between fraction and decimal forms of rational numbers.

Lesson Notes

Each student will need a calculator to complete this lesson.

Classwork

Example 1 (6 minutes): Can All Rational Numbers Be Written as Decimals?

- Can we find the decimal form of $\frac{1}{6}$ by writing it as an equivalent fraction with only factors of 2 or 5 in the denominator?
    - $\frac{1}{6} = \frac{1}{2 \times 3}$. *There are no factors of* 3 *in the numerator, so the factor of* 3 *has to remain in the denominator. This means we cannot write the denominator as a product of only* 2*'s and* 5*'s; therefore, the denominator cannot be a power of ten. The equivalent fraction method will not help us write* $\frac{1}{6}$ *as a decimal.*
- Is there another way to convert fractions to decimals?
    - *A fraction is interpreted as its numerator divided by its denominator. Since* $\frac{1}{6}$ *is a fraction, we can divide the numerator* 1 *by the denominator* 6*.*
- Use the division button on your calculator to divide 1 by 6.
- What do you notice about the quotient?
    - *It does not terminate, and almost all of the decimal places have the same number in them.*

Example 1: Can All Rational Numbers Be Written as Decimals?

a. **Using the division button on your calculator, explore various quotients of integers 1 through 11. Record your fraction representations and their corresponding decimal representations in the space below.**

Fractions will vary. Examples:

$\frac{1}{2} = 0.5$ $\frac{1}{3} = 0.3333333...$ $\frac{1}{4} = 0.25$ $\frac{1}{5} = 0.2$ $\frac{1}{6} = 0.1666666...$

$\frac{1}{7} = 0.1428571428...$ $\frac{1}{8} = 0.125$ $\frac{1}{9} = 0.1111111...$ $\frac{1}{10} = 0.1$ $\frac{1}{11} = 0.0909090909...$

b. **What two types of decimals do you see?**

Some of the decimals stop, and some fill up the calculator screen (or keep going).

- Did you find any quotients of integers that do not have decimal representations?
  - *No. Dividing by zero is not allowed. All quotients have decimal representations, but some do not terminate (end).*
- All rational numbers can be represented in the form of a decimal. We have seen already that fractions with powers of ten in their denominators (and their equivalent fractions) can be represented as terminating decimals. Therefore, other fractions must be represented by decimals that do not terminate.

Example 2 (4 minutes): Decimal Representations of Rational Numbers

Example 2: Decimal Representations of Rational Numbers

In the chart below, organize the fractions and their corresponding decimal representation listed in Example 1 according to their type of decimal.

Terminating	*Non-terminating*
$\frac{1}{2} = 0.5$	$\frac{1}{3} = 0.3333333\cdots$
$\frac{1}{4} = 0.25$	$\frac{1}{6} = 0.1666666\cdots$
$\frac{1}{5} = 0.2$	$\frac{1}{7} = 0.14285741258714\cdots$
$\frac{1}{8} = 0.125$	$\frac{1}{9} = 0.1111111\cdots$
$\frac{1}{10} = 0.1$	$\frac{1}{11} = 0.0909090909\cdots$
What do these fractions have in common?	**What do these fractions have in common?**
Each denominator is a product of only the factors 2 and/or 5.	*Each denominator contains at least one factor other than a 2 or a 5.*

EUREKA MATH

G7-M2-B2-v1.3.0_05/04/15

Example 3 (3 minutes): Converting Rational Numbers to Decimals Using Long Division

Part 1: Terminating Decimals

> *Scaffolding:*
> Review vocabulary of long division (i.e., algorithm, dividend, divisor, and remainder) for English language learners.

Example 3: Converting Rational Numbers to Decimals Using Long Division

Use the long division algorithm to find the decimal value of $-\frac{3}{4}$.

The fraction is a negative value, so its decimal representation will be as well.

$-\frac{3}{4} = -0.75$

```
    0.75
 4)3.00
  -2 8
     2 0
   - 2 0
       0
```

We know that $-\left(\frac{3}{4}\right) = -\frac{3}{4} = \frac{3}{-4}$, so we use our rules for dividing integers. Dividing 3 by 4 gives us 0.75, but we know the value must be negative.

Answer: -0.75

Exercise 1 (4 minutes)

> *Scaffolding:*
> For long division calculations, provide students with graph paper to aid with organization of numbers and decimal placement.

Exercise 1

Convert each rational number to its decimal form using long division.

a. $-\frac{7}{8} =$

```
    0.875
 8)7.000
  -6 4
     60
   - 56
     40
   - 40
      0
```

b. $\frac{3}{16} =$

$\frac{3}{16} = 0.1875$

```
     0.1875
 16)3.0000
   -1 6
    1 40
   -1 28
      120
    - 112
       80
     - 80
        0
```

Example 4 (5 minutes): Converting Rational Numbers to Decimals Using Long Division

Part 2: Repeating Decimals

MP.8

Example 4: Converting Rational Numbers to Decimals Using Long Division

Use long division to find the decimal representation of $\frac{1}{3}$.

The remainders repeat, yielding the same dividend remainder in each step. This repeating remainder causes the numbers in the quotient to repeat as well. Because of this pattern, the decimal will go on forever, so we cannot write the exact quotient.

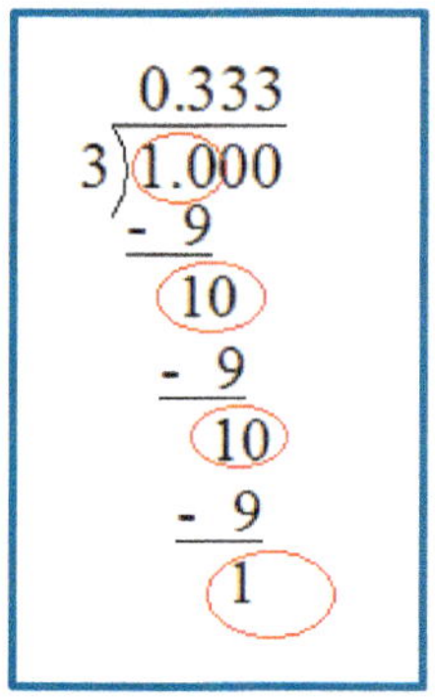

Students notice that since the remainders repeat, the quotient takes on a repeating pattern of 3s.

- We cannot possibly write the exact value of the decimal because it has an infinite number of decimal places. Instead, we indicate that the decimal has a repeating pattern by placing a bar over the shortest sequence of repeating digits (called the repetend).
  - $0.333\ldots = 0.\overline{3}$
- What part of your calculations causes the decimal to repeat?
  - *When a remainder repeats, the calculations that follow must also repeat in a cyclical pattern, causing the digits in the quotient to also repeat in a cyclical pattern.*
- Circle the repeating remainders.

Refer to the graphic above.

Exercise 2 (8 minutes)

Exercise 2

Calculate the decimal values of the fraction below using long division. Express your answers using bars over the shortest sequence of repeating digits.

a. $-\frac{4}{9}$

$-\frac{4}{9} = -0.4444\ldots = -0.\overline{4}$

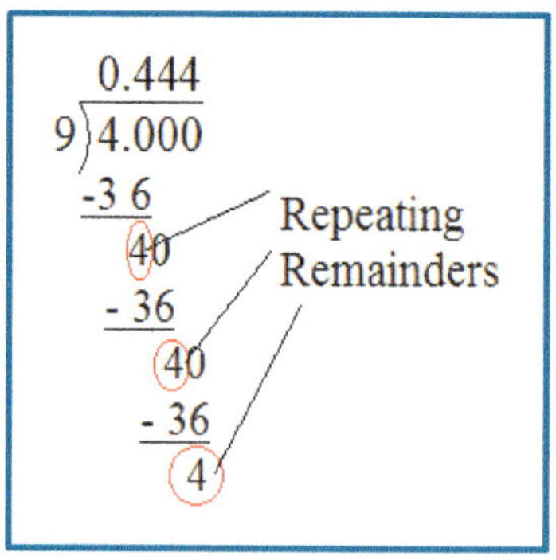

Scaffolding:

For long division calculations, provide students with graph paper to aid with organization of numbers and decimal placement.

b. $-\frac{1}{11}$

$-\frac{1}{11} = -0.090909... = -0.\overline{09}$

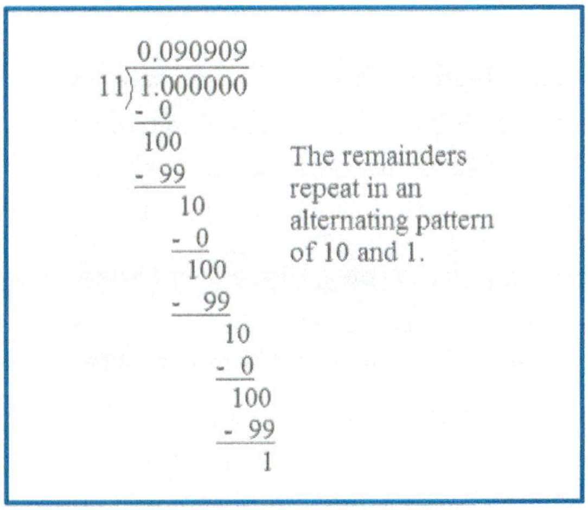

c. $\frac{1}{7}$

$\frac{1}{7} = 0.142857148... = 0.\overline{142857}$

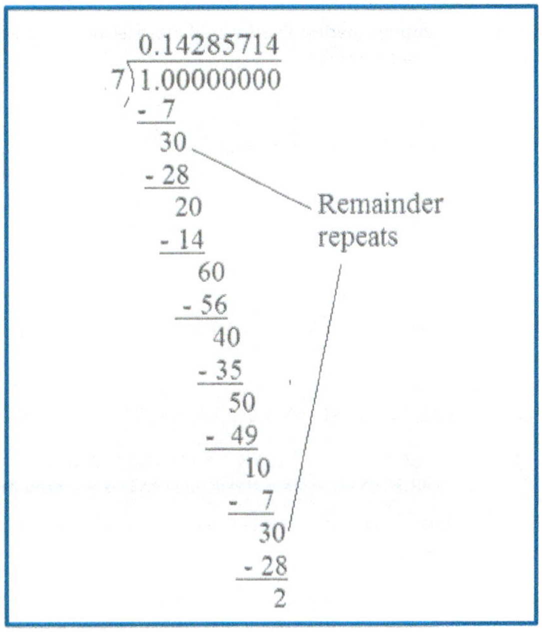

d. $-\frac{5}{6}$

$-\frac{5}{6} = -0.833333... = -0.8\overline{3}$

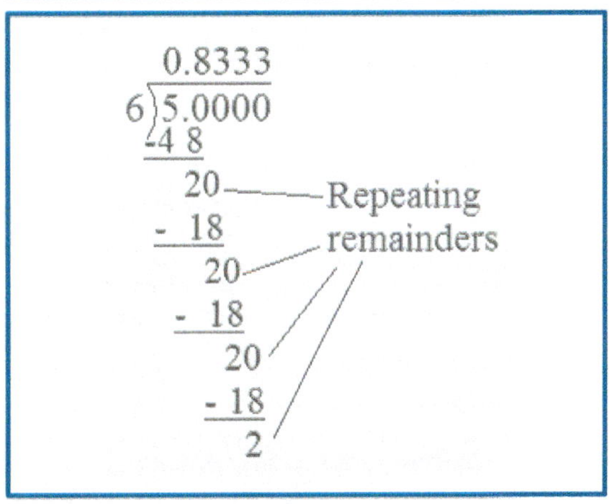

Example 5 (4 minutes): Fractions Represent Terminating or Repeating Decimals

- The long division algorithm either terminates with a zero remainder or the remainder repeats. Why?
  - Case 1: The long division algorithm terminates with a remainder of 0.
    - *The decimal also terminates.*
  - Case 2: The long division algorithm does not terminate with a remainder of 0.
- Consider $\frac{1}{7}$ from Exercise 2. There is no zero remainder, so the algorithm continues. The remainders cannot be greater than or equal to the divisor, 7, so there are only six possible nonzero remainders: 1, 2, 3, 4, 5, and 6. This means that the remainder must repeat within six steps.

MP.2 Students justify the claim in student materials.

Example 5: Fractions Represent Terminating or Repeating Decimals

How do we determine whether the decimal representation of a quotient of two integers, with the divisor not equal to zero, will terminate or repeat?

In the division algorithm, if the remainder is zero, then the algorithm terminates, resulting in a terminating decimal.

If the value of the remainder is not zero, then it is limited to whole numbers $1, 2, 3, \ldots, (d-1)$, where d is the divisor. This means that the value of the remainder must repeat within $(d-1)$ steps. (For example, given a divisor of 9, the nonzero remainders are limited to whole numbers 1 through 8, so the remainder must repeat within 8 steps.) When the remainder repeats, the calculations that follow will also repeat in a cyclical pattern causing a repeating decimal.

Example 6 (5 minutes): Using Rational Number Conversions in Problem Solving

Example 6: Using Rational Number Conversions in Problem Solving

a. **Eric and four of his friends are taking a trip across the New York State Thruway. They decide to split the cost of tolls equally. If the total cost of tolls is $\$8$, how much will each person have to pay?**

There are five people taking the trip. The friends will each be responsible for $\$1.60$ of the tolls due.

```
   1.6
5)8.0
 -5
  30
 -30
   0
```

b. **Just before leaving on the trip, two of Eric's friends have a family emergency and cannot go. What is each person's share of the $\$8$ tolls now?**

There are now three people taking the trip. The resulting quotient is a repeating decimal because the remainders repeat as 2s. The resulting quotient is $\frac{8}{3} = 2.66666\ldots = 2.\overline{6}$. If each friend pays $\$2.66$, they will be $\$0.02$ shy of $\$8$, so the amount must be rounded up to $\$2.67$ per person.

```
  1 1
  2.66
×    3
------
  7.98
not enough,
round up
```

```
   2.66
3)8.00
 -6
  2 0
 -1 8
    20
  - 18
     2
```

Closing (2 minutes)

MP.1

- What should you do if the remainders of a quotient of integers do not seem to repeat?
  - *Double check your work for computational errors, but if all is well, keep going! If you are doing the math correctly, the remainders eventually have to terminate or repeat.*
- What is the form for writing a repeating decimal?
  - *Use a bar to cover the shortest sequence of repeating digits.*

Present the following terminology:

REPEATING DECIMAL EXPANSION: A decimal expansion is *repeating* if, after some digit to the right of the decimal point, there is a finite string of consecutive digits called a block after which the decimal expansion consists entirely of consecutive copies of that block repeated forever.

A common notation for a repeating decimal expansion is to write out the decimal expansion until the end of the first block, and then write a line over the block.
For example, $3.1\overline{25}$ is a compact way to write the repeating decimal expansion $3.125252525\ldots$.

TERMINATING DECIMAL EXPANSION: A *terminating decimal expansion* is a repeating decimal expansion with period 1 and repeating digit 0.
For example, the terminating decimal expansion of $\frac{1}{4}$ is $0.25000\ldots$, or $0.25\overline{0}$.

A number with a terminating decimal expansion can be written as a finite decimal.
For example, the terminating decimal expansion $0.25000\ldots$ represents the same number as the finite decimal 0.25.

> Lesson Summary
>
> **The real world requires that we represent rational numbers in different ways depending on the context of a situation. All rational numbers can be represented as either terminating decimals or repeating decimals using the long division algorithm. We represent repeating decimals by placing a bar over the shortest sequence of repeating digits.**

Exit Ticket (4 minutes)

Name ____ Date ________________

Lesson 14: Converting Rational Numbers to Decimals Using Long Division

Exit Ticket

1. What is the decimal value of $\frac{4}{11}$?

2. How do you know that $\frac{4}{11}$ is a repeating decimal?

3. What causes a repeating decimal in the long division algorithm?

Exit Ticket Sample Solutions

1. What is the decimal value of $\frac{4}{11}$?

   $\frac{4}{11} = 0.\overline{36}$

2. How do you know that $\frac{4}{11}$ is a repeating decimal?

   The prime factor in the denominator is 11. Fractions that correspond with terminating decimals have only factors 2 and 5 in the denominator in simplest form.

3. What causes a repeating decimal in the long division algorithm?

   When a remainder repeats, the division algorithm takes on a cyclic pattern causing a repeating decimal.

Problem Set Sample Solutions

1. Convert each rational number into its decimal form.

$\frac{1}{9} = 0.\overline{1}$

$\frac{1}{6} = 0.1\overline{6}$

$\frac{2}{9} = 0.\overline{2}$

$\frac{1}{3} = 0.\overline{3}$ $\quad$ $\frac{2}{6} = 0.\overline{3}$ $\quad$ $\frac{3}{9} = 0.\overline{3}$

$\frac{4}{9} = 0.\overline{4}$

$\frac{3}{6} = 0.5$

$\frac{5}{9} = 0.\overline{5}$

$\frac{2}{3} = 0.\overline{6}$ $\quad$ $\frac{4}{6} = 0.\overline{6}$ $\quad$ $\frac{6}{9} = 0.\overline{6}$

$\frac{7}{9} = 0.\overline{7}$

$\frac{5}{6} = 0.8\overline{3}$

$\frac{8}{9} = 0.\overline{8}$

One of these decimal representations is not like the others. Why?

$\frac{3}{6}$ in its simplest form is $\frac{1}{2}$ (the common factor of 3 divides out, leaving a denominator of 2, which in decimal form will terminate.

Enrichment:

2. Chandler tells Aubrey that the decimal value of $-\frac{1}{17}$ is not a repeating decimal. Should Aubrey believe him? Explain.

   No, Aubrey should not believe Chandler. The divisor 17 is a prime number containing no factors of 2 or 5, and therefore, cannot be written as a terminating decimal. By long division, $-\frac{1}{17} = -0.\overline{0588235294117647}$. The decimal appears as though it is not going to take on a repeating pattern because all 16 possible nonzero remainders appear before the remainder repeats. The seventeenth step produces a repeat remainder causing a cyclical decimal pattern.

3. Complete the quotients below without using a calculator, and answer the questions that follow.

   a. Convert each rational number in the table to its decimal equivalent.

$\frac{1}{11} = 0.\overline{09}$	$\frac{2}{11} = 0.\overline{18}$	$\frac{3}{11} = 0.\overline{27}$	$\frac{4}{11} = 0.\overline{36}$	$\frac{5}{11} = 0.\overline{45}$
$\frac{6}{11} = 0.\overline{54}$	$\frac{7}{11} = 0.\overline{63}$	$\frac{8}{11} = 0.\overline{72}$	$\frac{9}{11} = 0.\overline{81}$	$\frac{10}{11} = 0.\overline{90}$

   Do you see a pattern? Explain.

   The two digits that repeat in each case have a sum of nine. As the numerator increases by one, the first of the two digits increases by one as the second of the digits decreases by one.

   b. Convert each rational number in the table to its decimal equivalent.

$\frac{0}{99} = 0$	$\frac{10}{99} = 0.\overline{10}$	$\frac{20}{99} = 0.\overline{20}$	$\frac{30}{99} = 0.\overline{30}$	$\frac{45}{99} = 0.\overline{45}$
$\frac{58}{99} = 0.\overline{58}$	$\frac{62}{99} = 0.\overline{62}$	$\frac{77}{99} = 0.\overline{7}$	$\frac{81}{99} = 0.\overline{81}$	$\frac{98}{99} = 0.\overline{98}$

   Do you see a pattern? Explain.

   The 2-digit numerator in each fraction is the repeating pattern in the decimal form.

   c. Can you find other rational numbers that follow similar patterns?

   Answers will vary.

Lesson 15: Multiplication and Division of Rational Numbers

Student Outcomes

- Students recognize that the rules for multiplying and dividing integers apply to rational numbers.
- Students interpret products and quotients of rational numbers by describing real-world contexts.

Classwork

Fluency Exercise (6 minutes): Integer Multiplication

Photocopy the attached 2-page fluency-building exercises so that each student receives a copy. Before beginning, tell students that they may not skip over questions and that they must move in order. Allow students one minute to complete Side A. After one minute, discuss the answers. Before beginning Side B, elicit strategies from those students who were able to accurately complete the most problems on Side A. Administer Side B in the same fashion, and review the answers. Refer to the Sprints and Sprint Delivery Script sections in the Module Overview for directions to administer a Sprint.

Exercise 1 (6 minutes)

MP.2 Students work for two minutes with learning partners or a group to create a word problem involving integer multiplication. Students may use white boards or a half sheet of paper to record the word problem. Every group member should record the word problem and its answer in his student materials.

> *Scaffolding:*
> For students who are not yet fluent with integer multiplication, provide cards with the rules for integer multiplication.

For the next three minutes, groups switch work (white boards or half sheets) and solve the word problem they receive. Students verify that the problem can be solved using multiplication of integers. Once students solve their problem, they check back with the group who created it to make sure they are in agreement on the answer.

For the remaining two minutes, students take their original word problem and modify it in their student materials by replacing an integer with another signed number that is either a fraction or decimal. Students rework the problem and arrive at the answer to the new problem, recording their work in their student materials.

Exercise 1

a. **In the space below, create a word problem that involves integer multiplication. Write an equation to model the situation.**

Answers will vary.

Both times we went to the fair, I borrowed $\$3$ *from my older brother.* $-\$3 \times 2 = -\$6$

b. **Now change the word problem by replacing the integers with non-integer rational numbers (fractions or decimals), and write the new equation.**

Answers will vary.

Both times we went to the fair, I borrowed $\$3.50$ *from my older brother.* $-\$3.50 \times 2 = -\$7.00$

c. **Was the process used to solve the second problem different from the process used to solve the first? Explain.**

No, the process was the same. Both times I had a positive number multiplied by a negative number, so the product is a negative number. The process, multiplication, is represented as repeated addition: $-\$3.50 + (-\$3.50) = -\$7.00$.

- Was the process for solving the second problem different from the process you used to solve the problem when it contained only integers?
  - *No*

Students record the rules in Exercise 1, part (d) of their student materials.

d. **The Rules for Multiplying Rational Numbers are the same as the Rules for Multiplying Integers:**

1. *Multiply the absolute values of the two rational numbers.*
2. *If the two numbers (factors) have the same sign, their product is positive.*
3. *If the two numbers (factors) have opposite signs, their product is negative.*

Exercise 2 (5 minutes)

Students work independently to answer the following question in their student materials. They write an equation involving rational numbers and show all computational work. Students discuss their long division work with their learning partners until they agree on the answer.

Exercise 2

a. **In one year, Melinda's parents spend $\$2,640.90$ on cable and internet service. If they spend the same amount each month, what is the resulting monthly change in the family's income?**

$-2,640.90 \div 12 = -220.08$

The average monthly change to their income is about $-\$220.08$.

```
    220.075
12)2640.900
  -24
    24
   -24
     0 09
    -  0
       90
     - 84
        60
      - 60
         0
```

- Are the rules for dividing rational numbers the same as the rules for dividing integers?
  - *Yes*

Students record the rules in Exercise 2, part (b) of their student materials.

b. **The Rules for Dividing Rational Numbers are the same as the Rules for Dividing Integers:**

1. *Divide the absolute values of the two rational numbers.*
2. *If the two numbers (dividend and divisor) have the same sign, their quotient is positive.*
3. *If the two numbers (dividend and divisor) have opposite signs, their quotient is negative.*

Exercise 3 (20 minutes)

Exercise 3

Use the fundraiser chart to help answer the questions that follow.

Grimes Middle School Flower Fundraiser

Customer	Plant Type	Number of Plants	Price per Plant	Total	Paid? Yes or No
Tamara Jones	tulip	2	$\$4.25$	$\$8.50$	No
Mrs. Wolff	daisy	1	$\$3.75$	$\$3.75$	Yes
Mr. Clark	geranium	5	$\$2.25$	$\$11.25$	Yes
Susie (Jeremy's sister)	violet	1	$\$2.50$	$\$2.50$	Yes
Nana and Pop (Jeremy's grandparents)	daisy	4	$\$3.75$	$\$15.00$	No

Jeremy is selling plants for the school's fundraiser, and listed above is a chart from his fundraiser order form. Use the information in the chart to answer the following questions. Show your work, and represent the answer as a rational number; then, explain your answer in the context of the situation.

a. **If Tamara Jones writes a check to pay for the plants, what is the resulting change in her checking account balance?**

$-4.25 \times 2 = -8.50$

Numerical Answer: -8.50

Explanation: *Tamara Jones will need to deduct* $\$8.50$ *from her checking account balance.*

b. **Mr. Clark wants to pay for his order with a $\$20$ bill, but Jeremy does not have change. Jeremy tells Mr. Clark he will give him the change later. How will this affect the total amount of money Jeremy collects? Explain. What rational number represents the change that must be made to the money Jeremy collects?**

$2.25 \times 5 = 11.25$ $20.00 - 11.25 = 8.75$

Numerical Answer: -8.75

Explanation: *Jeremy collects too much money. He owes Mr. Clark* $\$8.75$. *The adjustment Jeremy needs to make is* $-\$8.75$.

c. **Jeremy's sister, Susie, borrowed the money from their mom to pay for her order. Their mother has agreed to deduct an equal amount of money from Susie's allowance each week for the next five weeks to repay the loan. What is the weekly change in Susie's allowance?**

$-2.50 \div 5 = -0.50$

Numerical Answer: -0.50

Explanation: *Susie will lose* $\$0.50$ *of her allowance each week.*

d. **Jeremy's grandparents want to change their order. They want to order three daisies and one geranium, instead of four daisies. How does this change affect the amount of their order? Explain how you arrived at your answer.**

Original Order: $3.75 \times 4 = 15.00$

New Order: $3.75 \times 3 + 2.25 = 11.25 + 2.25 = 13.50$

$15.00 - 13.50 = 1.50$

Numerical Answer: 1.50

Explanation: Jeremy's grandparents will get back $\$1.50$ *since the change in their order made it cheaper. I got my answer by first calculating the cost of the original order. Second, I calculated the cost of the new order. Finally, I subtracted the two order totals to determine the difference in cost.*

e. **Jeremy approaches three people who do not want to buy any plants; however, they wish to donate some money for the fundraiser when Jeremy delivers the plants one week later. If the people promise to donate a total of $\$14.40$, what will be the average cash donation?**

$14.40 \div 3 = 4.80$

Numerical Answer: 4.80

Explanation: The average cash donation will be $\$4.80$ *per person.*

f. **Jeremy spends one week collecting orders. If 22 people purchase plants totaling $\$270$, what is the average amount of Jeremy's sale?**

$270 \div 22 \approx 12.272$

Numerical Answer: 12.27

Explanation: The average sale is about $\$12.27$.

```
      12.272
22)270.000
   -22
     50
   - 44
      60
    - 44
      160
    - 154
        60
      - 44
        16
```

Closing (2 minutes)

MP.7

- When answering word problems today about the Grimes Middle School Flower Fundraiser, how did you know whether to multiply or divide?
  - *Answers will vary.*
- How did you know whether to express your answer as a positive or negative number?
  - *Answers will vary. Encourage students to refer back to the rules of multiplication and division of rational numbers.*
- In general, how does the context of a word problem indicate whether you should multiply or divide rational numbers and how your answer will be stated?
  - *When reading word problems, we can look for key words that indicate multiplication or division. There are also key words that tell us if a value is positive or negative.*

Lesson Summary

The rules that apply for multiplying and dividing integers apply to rational numbers. We can use the products and quotients of rational numbers to describe real-world situations.

Exit Ticket (6 minutes)

Name ______________________________ Date ______________

Lesson 15: Multiplication and Division of Rational Numbers

Exit Ticket

Harrison made up a game for his math project. It is similar to the Integer Game; however, in addition to integers, there are cards that contain other rational numbers such as -0.5 and -0.25. Write a multiplication or division equation to represent each problem below. Show all related work.

1. Harrison discards three -0.25 cards from his hand. How does this affect the overall point value of his hand? Write an equation to model this situation.

2. Ezra and Benji are playing the game with Harrison. After Ezra doubles his hand's value, he has a total of -14.5 points. What was his hand's value before he doubled it?

3. Benji has four -0.5 cards. What is his total score?

Exit Ticket Sample Solutions

Harrison made up a game for his math project. It is similar to the Integer Game; however, in addition to integers, there are cards that contain other rational numbers such as -0.5 and -0.25. Write a multiplication or division equation to represent each problem below. Show all related work.

1. **Harrison discards three -0.25 cards from his hand. How does this affect the overall point value of his hand? Write an equation to model this situation.**

   $-3\,(-0.25) = 0.75$

   The point value of Harrison's hand will increase by 0.75 points.

2. **Ezra and Benji are playing the game with Harrison. After Ezra doubles his hand's value, he has a total of -14.5 points. What was his hand's value before he doubled it?**

   $-14.5 \div 2 = -7.25$

   Before Ezra doubled his hand, his hand had a point value of -7.25.

3. **Benji has four -0.5 cards. What is his total score?**

   $4 \times (-0.5) = -2.0$

   Benji's total score is -2.0 points.

Problem Set Sample Solutions

1. **At lunch time, Benjamin often borrows money from his friends to buy snacks in the school cafeteria. Benjamin borrowed $\$0.75$ from his friend Clyde five days last week to buy ice cream bars. Represent the amount Benjamin borrowed as the product of two rational numbers; then, determine how much Benjamin owed his friend last week.**

   $5\,(-0.75) = -3.75$

   Benjamin owed Clyde $\$3.75$.

2. **Monica regularly records her favorite television show. Each episode of the show requires 3.5% of the total capacity of her video recorder. Her recorder currently has 62% of its total memory free. If Monica records all five episodes this week, how much space will be left on her video recorder?**

   $62 + 5(-3.5) = 62 + (-17.5) = 44.5$

   Monica's recorder will have 44.5% of its total memory free.

For Problems 3–5, find at least two possible sets of values that will work for each problem.

3. **Fill in the blanks with two rational numbers (other than 1 and -1).** $\_\_\_ \times \left(-\frac{1}{2}\right) \times \_\_\_ = -20$

   What must be true about the relationship between the two numbers you chose?

   Answers may vary. Two possible solutions are 10 and 4 or -10 and -4. The two numbers must be factors of 40, and they must both have the same sign.

4. Fill in the blanks with two rational numbers (other than 1 and -1). $-5.6 \times 100 \div 80 \times$ ___ $\times$ ___ $= 700$

What must be true about the relationship between the two numbers you chose?

Answers may vary. Two possible solutions are -50 and 2 or 25 and -4. The two numbers must be factors of -100, and they must have opposite signs.

5. Fill in the blanks with two rational numbers. ___ $\times$ ___ $= -0.75$

What must be true about the relationship between the two numbers you chose?

Answers may vary. Two possible solutions are -3 and 0.25 or 0.5 and -1.5. The two numbers must be factors of -0.75, and they must have opposite signs.

For Problems 6–8, create word problems that can be represented by each expression, and then represent each product or quotient as a single rational number.

6. $8 \times (-0.25)$

Answers may vary.

Example: Stacey borrowed a quarter from her mother every time she went to the grocery store so that she could buy a gumball from the gumball machine. Over the summer, Stacey went to the grocery store with her mom eight times. What rational number represents the dollar amount change in her mother's money due to the purchase of gumballs?

Answer: -2

7. $-6 \div \left(1\frac{1}{3}\right)$

Answers may vary.

Example: There was a loss of $\$6$ on my investment over one-and-a-third months. Based on this, what was the investment's average dollar amount change per month?

Answer: -4.50

8. $-\frac{1}{2} \times 12$

Answers may vary.

Example: I discarded exactly half of my card-point total in the Integer Game. If my card-point total was 12 before I discarded, which rational number represents the change to my hand's total card-point total?

Answer: -6

Number Correct: ______

Integer Multiplication—Round 1

Directions: Determine the product of the integers, and write it in the column to the right.

1.	$-2 \cdot -8$		23.	$-14 \cdot -12$	
2.	$-4 \cdot 3$		24.	$15 \cdot -13$	
3.	$5 \cdot -7$		25.	$16 \cdot -18$	
4.	$1 \cdot -1$		26.	$24 \cdot -17$	
5.	$-6 \cdot 9$		27.	$-32 \cdot -21$	
6.	$-2 \cdot -7$		28.	$19 \cdot -27$	
7.	$8 \cdot -3$		29.	$-39 \cdot 10$	
8.	$0 \cdot -9$		30.	$43 \cdot 22$	
9.	$12 \cdot -5$		31.	$11 \cdot -33$	
10.	$-4 \cdot 2$		32.	$-29 \cdot -45$	
11.	$-1 \cdot -6$		33.	$37 \cdot -44$	
12.	$10 \cdot -4$		34.	$-87 \cdot -100$	
13.	$14 \cdot -3$		35.	$92 \cdot -232$	
14.	$-5 \cdot -13$		36.	$456 \cdot 87$	
15.	$-16 \cdot -8$		37.	$-143 \cdot 76$	
16.	$18 \cdot -2$		38.	$439 \cdot -871$	
17.	$-15 \cdot 7$		39.	$-286 \cdot -412$	
18.	$-19 \cdot 1$		40.	$-971 \cdot 342$	
19.	$12 \cdot 12$		41.	$-773 \cdot -407$	
20.	$9 \cdot -17$		42.	$-820 \cdot 638$	
21.	$-8 \cdot -14$		43.	$591 \cdot -734$	
22.	$-7 \cdot 13$		44.	$491 \cdot -197$	

Integer Multiplication—Round 1 [KEY]

Directions: Determine the product of the integers, and write it in the column to the right.

1.	$-2 \cdot -8$	16	23.	$-14 \cdot -12$	168
2.	$-4 \cdot 3$	-12	24.	$15 \cdot -13$	-195
3.	$5 \cdot -7$	-35	25.	$16 \cdot -18$	-288
4.	$1 \cdot -1$	-1	26.	$24 \cdot -17$	-408
5.	$-6 \cdot 9$	-54	27.	$-32 \cdot -21$	672
6.	$-2 \cdot -7$	14	28.	$19 \cdot -27$	-513
7.	$8 \cdot -3$	-24	29.	$-39 \cdot 10$	-390
8.	$0 \cdot -9$	0	30.	$43 \cdot 22$	946
9.	$12 \cdot -5$	-60	31.	$11 \cdot -33$	-363
10.	$-4 \cdot 2$	-8	32.	$-29 \cdot -45$	$1,305$
11.	$-1 \cdot -6$	6	33.	$37 \cdot -44$	$-1,628$
12.	$10 \cdot -4$	-40	34.	$-87 \cdot -100$	$8,700$
13.	$14 \cdot -3$	-42	35.	$92 \cdot -232$	$-21,344$
14.	$-5 \cdot -13$	65	36.	$456 \cdot 87$	$39,672$
15.	$-16 \cdot -8$	128	37.	$-143 \cdot 76$	$-10,868$
16.	$18 \cdot -2$	-36	38.	$439 \cdot -871$	$-382,369$
17.	$-15 \cdot 7$	-105	39.	$-286 \cdot -412$	$117,832$
18.	$-19 \cdot 1$	-19	40.	$-971 \cdot 342$	$-332,082$
19.	$12 \cdot 12$	144	41.	$-773 \cdot -407$	$314,611$
20.	$9 \cdot -17$	-153	42.	$-820 \cdot 638$	$-523,160$
21.	$-8 \cdot -14$	112	43.	$591 \cdot -734$	$-433,794$
22.	$-7 \cdot 13$	-91	44.	$491 \cdot -197$	$-96,727$

Number Correct: ______

Improvement: ______

Integer Multiplication—Round 2

Directions: Determine the product of the integers, and write it in the column to the right.

1.	$-9 \cdot -7$		23.	$-22 \cdot 14$	
2.	$0 \cdot -4$		24.	$-18 \cdot -32$	
3.	$3 \cdot -5$		25.	$-24 \cdot 19$	
4.	$6 \cdot -8$		26.	$47 \cdot 21$	
5.	$-2 \cdot 1$		27.	$17 \cdot -39$	
6.	$-6 \cdot 5$		28.	$-16 \cdot -28$	
7.	$-10 \cdot -12$		29.	$-67 \cdot -81$	
8.	$11 \cdot -4$		30.	$-36 \cdot 44$	
9.	$3 \cdot 8$		31.	$-50 \cdot 23$	
10.	$12 \cdot -7$		32.	$66 \cdot -71$	
11.	$-1 \cdot 8$		33.	$82 \cdot -29$	
12.	$5 \cdot -10$		34.	$-32 \cdot 231$	
13.	$3 \cdot -13$		35.	$89 \cdot -744$	
14.	$15 \cdot -8$		36.	$623 \cdot -22$	
15.	$-9 \cdot 14$		37.	$-870 \cdot -46$	
16.	$-17 \cdot 5$		38.	$179 \cdot 329$	
17.	$16 \cdot 2$		39.	$-956 \cdot 723$	
18.	$19 \cdot -7$		40.	$874 \cdot -333$	
19.	$-6 \cdot 13$		41.	$908 \cdot -471$	
20.	$1 \cdot -18$		42.	$-661 \cdot -403$	
21.	$-14 \cdot -3$		43.	$-520 \cdot -614$	
22.	$-10 \cdot -17$		44.	$-309 \cdot 911$	

Integer Multiplication—Round 2 [KEY]

Directions: Determine the product of the integers, and write it in the column to the right.

1.	$-9 \cdot -7$	63	23.	$-22 \cdot 14$	-308
2.	$0 \cdot -4$	0	24.	$-18 \cdot -32$	576
3.	$3 \cdot -5$	-15	25.	$-24 \cdot 19$	-456
4.	$6 \cdot -8$	-48	26.	$47 \cdot 21$	987
5.	$-2 \cdot 1$	-2	27.	$17 \cdot -39$	-663
6.	$-6 \cdot 5$	-30	28.	$-16 \cdot -28$	448
7.	$-10 \cdot -12$	120	29.	$-67 \cdot -81$	$5,427$
8.	$11 \cdot -4$	-44	30.	$-36 \cdot 44$	$-1,584$
9.	$3 \cdot 8$	24	31.	$-50 \cdot 23$	$-1,150$
10.	$12 \cdot -7$	-84	32.	$66 \cdot -71$	$-4,686$
11.	$-1 \cdot 8$	-8	33.	$82 \cdot -29$	$-2,378$
12.	$5 \cdot -10$	-50	34.	$-32 \cdot 231$	$-7,392$
13.	$3 \cdot -13$	-39	35.	$89 \cdot -744$	$66,216$
14.	$15 \cdot -8$	-120	36.	$623 \cdot -22$	$-13,706$
15.	$-9 \cdot 14$	-126	37.	$-870 \cdot -46$	$40,020$
16.	$-17 \cdot 5$	-85	38.	$179 \cdot 329$	$58,891$
17.	$16 \cdot 2$	32	39.	$-956 \cdot 723$	$-691,188$
18.	$19 \cdot -7$	-133	40.	$874 \cdot -333$	$-291,042$
19.	$-6 \cdot 13$	-78	41.	$908 \cdot -471$	$-427,668$
20.	$1 \cdot -18$	-18	42.	$-661 \cdot -403$	$266,383$
21.	$-14 \cdot -3$	42	43.	$-520 \cdot -614$	$319,280$
22.	$-10 \cdot -17$	170	44.	$-309 \cdot 911$	$-281,499$

Lesson 16: Applying the Properties of Operations to Multiply and Divide Rational Numbers

Student Outcomes

- Students use properties of operations to multiply and divide rational numbers without the use of a calculator. They use the commutative and associative properties of multiplication to generate equivalent expressions. They use the distributive property of multiplication over addition to create equivalent expressions, representing the sum of two quantities with a common factor as a product, and vice-versa.
- Students recognize that any problem involving multiplication and division can be written as a problem involving only multiplication.
- Students determine the sign of an expression that contains products and quotients by checking whether the number of negative terms is even or odd.

Lesson Notes

Recall the Properties of Operations from Table 3 of the Common Core State Standards.

Table 3. The Properties of Operations. Here $a, b,$ and c stand for arbitrary numbers.

Associative property of addition	$(a+b)+c=a+(b+c)$
Commutative property of addition	$a+b=b+a$
Additive identity property of 0	$a+0=0+a=a$
Existence of additive inverses	For every a there exists $-a$ so that $a+(-a)=(-a)+a=0$
Associative property of multiplication	$(a\cdot b)\cdot c=a\cdot(b\cdot c)$
Commutative property of multiplication	$a\cdot b=b\cdot a$
Multiplicative identity property of 1	$a\cdot 1=1\cdot a=a$
Existence of multiplicative inverses	For every $a\neq 0$ there exists $\frac{1}{a}$ so that $a\cdot\frac{1}{a}=\frac{1}{a}\cdot a=1$
Distributive property of multiplication over addition	$a\cdot(b+c)=a\cdot b+a\cdot c$

Classwork

Fluency Exercise (6 minutes): Integer Division

Photocopy the attached 2-page fluency-building exercises so that each student receives a copy. Before beginning, inform students that they may not skip over questions and that they must move in order. Allow students one minute to complete Side A. After one minute, discuss the answers. Before moving on to Side B, elicit strategies from those students who were able to accurately complete many problems on Side A. Administer Side B in the same fashion, and review the answers. Refer to the Sprints and Sprint Delivery Script sections in the Module Overview for directions to administer a Sprint.

Example 1 (5 minutes): Using the Commutative and Associative Properties to Efficiently Multiply Rational Numbers

Present the question below and have students share their thoughts. Answers will vary, but accept all answers at this point.

- How can we evaluate the expression below? Will different strategies result in different answers? Why or why not?

$$-6 \times 2 \times (-2) \times (-5) \times (-3)$$

Example 1: Using the Commutative and Associative Properties to Efficiently Multiply Rational Numbers

a. **Evaluate the expression below.**

$$-6 \times 2 \times (-2) \times (-5) \times (-3)$$

$-6 \times 2 \times (-2) \times (-5) \times (-3)$	$-6 \times 2 \times (-2) \times (-5) \times (-3)$	
$-12 \times (-2) \times (-5) \times (-3)$	$-6 \times 2 \times 10 \times (-3)$	*Associative property of multiplication*
$24 \times (-5) \times (-3)$	$-6 \times 2 \times (-3) \times 10$	*Commutative property of multiplication*
$-120 \times (-3)$	$-6 \times (-6) \times 10$	*Associative property of multiplication*
360	36×10	
	360	

Students experiment with different strategies from their discussion to evaluate the product of integers. After time to work, student groups share their strategies and solutions. Students and teacher discuss the properties (commutative and associative) that allow us to manipulate expressions.

b. **What types of strategies were used to evaluate the expressions?**

The strategies used were order of operations, rearranging the terms using the commutative property, and multiplying the terms in various orders using the associative property.

c. **Can you identify the benefits of choosing one strategy versus another?**

Multiplying the terms allowed me to combine factors in more manageable ways, such as multiplying $(-2) \times (-5)$ *to get* 10*. Multiplying other numbers by* 10 *is very easy.*

d. What is the sign of the product, and how was the sign determined?

The product is a positive value. When calculating the product of the first two factors, the answer will be negative because when the factors have opposite signs, the result is a negative product. Two negative values multiplied together yield a positive product. When a negative value is multiplied by a positive product, the sign of the product again changes to a negative value. When this negative product is multiplied by the last (fourth) negative value, the sign of the product again changes to a positive value.

Exercise 1 (3 minutes)

Exercise 1

Find an efficient strategy to evaluate the expression and complete the necessary work.

Methods will vary.

$-1 \times (-3) \times 10 \times \underbrace{(-2) \times 2}$ *Associative property*

$\underbrace{-1 \times (-3)} \times 10 \times (-4)$

$3 \times 10 \times (-4)$

$\underbrace{3 \times (-4)} \times 10$ *Commutative multiplication*

$\underbrace{-12 \times 10}$

-120

Discussion

- What aspects of the expression did you consider when choosing a strategy for evaluating this expression?
  - *Answers will vary.*
- What is the sign of the product, and how was the sign determined?
  - *The sign of the product is negative. If we follow the method above, the first two factors result in a negative product because they have opposite signs. The next two factors we used to calculate the product were both negative, so the result was a positive product. The next product we calculated was negative because the factors had opposite signs. Finally, the final two factors also had opposite signs, so the final product was negative.*
- How else could we have evaluated this problem?
  - *Answers may vary, but two possible answers are provided.*
  - *We could have solved this problem by following order of operations or multiplying the factors left to right.*

Exercises 2–4 (6 minutes)

- Is order of operations an efficient strategy to multiply the expression below? Why or why not?

$$4 \times \frac{1}{3} \times (-8) \times 9 \times \left(-\frac{1}{2}\right)$$

After discussion, student groups choose a strategy to evaluate the expression:

Exercise 2

Find an efficient strategy to evaluate the expression and complete the necessary work.

Methods will vary.

$$4 \times \frac{1}{3} \times (-8) \times 9 \times \left(-\frac{1}{2}\right)$$

$$4 \times \underbrace{\left[\left(-\frac{1}{2}\right) \times (-8)\right]} \times 9 \times \frac{1}{3}$$ *Commutative multiplication*

$$4 \times \quad 4 \times \underbrace{\left[9 \times \frac{1}{3}\right]}$$ *Associative property*

$$4 \times \quad \underbrace{4 \times [3]}$$ *Associative property*

$$\underbrace{4 \times \quad 12}$$ *Associative property*

$$48$$

Exercise 3

What terms did you combine first and why?

I multiplied the $-\frac{1}{2} \times -8$ *and* $\frac{1}{3} \times 9$ *because their products are integers; this eliminated the fractions.*

MP.8

Exercise 4

Refer to the example and exercises. Do you see an easy way to determine the sign of the product first?

The product of two negative integers yields a positive product. If there is an even number of negative factors, then each negative value can be paired with another negative value yielding a positive product. This means that all factors become positive values and, therefore, have a positive product.

For example: $\underbrace{(-1) \times (-1)}_{1} \times \underbrace{(-1) \times (-1)}_{1} \times \underbrace{(-1) \times (-1)}_{1} = 1$

If there are an odd number of negative factors, then all except one can be paired with another negative. This leaves us with a product of a positive value and a negative value, which is negative.

For example: $\underbrace{(-1) \times (-1)}_{1} \times \underbrace{(-1) \times (-1)}_{1} \times \underbrace{(-1) \times (-1)}_{1} \times (-1)$

$$1 \times \underbrace{1 \times 1 \times (-1)}$$

$$1 \times (-1) = -1$$

Example 2 (4 minutes): Using the Distributive Property to Multiply Rational Numbers

- What is a mixed number?
  - *A mixed number is the sum of a whole number and a fraction.*
- What does the opposite of a mixed number look like?
  - *The opposite of a sum is equal to the sum of its opposites.*

> *Scaffolding:*
>
> Remind students that *the opposite of a sum is equivalent to the sum of its opposites.*

Example 2: Using the Distributive Property to Multiply Rational Numbers

Rewrite the mixed number as a sum; then, multiply using the distributive property.

$-6 \times \left(5\frac{1}{3}\right)$

$-6 \times \left(5 + \frac{1}{3}\right)$

$\underbrace{(-6 \times 5)}_{-30} + \underbrace{\left(-6 \times \frac{1}{3}\right)}_{(-2)}$ *Distributive property*

$-30 + (-2)$

-32

- Did the distributive property make this problem easier to evaluate? How so?
  - *Answers will vary, but most students will think that distributive property did make the problem easier to solve.*

Exercise 5 (3 minutes)

Exercise 5

Multiply the expression using the distributive property.

$9 \times \left(-3\frac{1}{2}\right)$

$9 \times \left(-3 + \left(-\frac{1}{2}\right)\right)$

$\underbrace{(9 \times (-3))}_{-27} + \underbrace{\left(9 \times \left(-\frac{1}{2}\right)\right)}_{\left(-4\frac{1}{2}\right)}$

$-27 + \left(-4\frac{1}{2}\right)$

$-31\frac{1}{2}$

Example 3 (4 minutes): Using the Distributive Property to Multiply Rational Numbers

The teacher and students together complete the given expression with justification.

Example 3: Using the Distributive Property to Multiply Rational Numbers

Evaluate using the distributive property.

$16 \times \left(-\frac{3}{8}\right) + 16 \times \frac{1}{4}$

$16\left(-\frac{3}{8} + \frac{1}{4}\right)$ *Distributive property*

$16\left(-\frac{3}{8} + \frac{2}{8}\right)$ *Equivalent fractions*

$16\left(-\frac{1}{8}\right)$

-2

Example 4 (4 minutes): Using the Multiplicative Inverse to Rewrite Division as Multiplication

- How is this expression different from the previous examples, and what can we do to make it more manageable?
  - *This expression involves division by fractions, and we know that dividing by a number is equivalent to multiplying by its multiplicative inverse (reciprocal); so, we can rewrite the entire expression as multiplication.*

Example 4: Using the Multiplicative Inverse to Rewrite Division as Multiplication

Rewrite the expression as only multiplication and evaluate.

$1 \div \frac{2}{3} \times (-8) \times 3 \div \left(-\frac{1}{2}\right)$

$1 \times \frac{3}{2} \times (-8) \times 3 \times (-2)$ — *Multiplicative inverse*

$1 \times \left[(-2) \times \left(\frac{3}{2}\right)\right] \times (-8) \times 3$ — *Commutative multiplication*

$1 \times [-3] \times (-8) \times 3$ — *Associative property*

$-3 \times (-8) \times 3$

$-3 \times 3 \times (-8)$ — *Commutative multiplication*

$-9 \times (-8)$

72

Exercise 6 (4 minutes)

Students in groups evaluate the following expression using the multiplicative inverse property. Methods will vary.

Exercise 6

$4.2 \times \left(-\frac{1}{3}\right) \div \frac{1}{6} \times (-10)$

$4.2 \times \left(-\frac{1}{3}\right) \times \frac{6}{1} \times (-10)$ — *Multiplicative inverse*

$4.2 \times (-10) \times \left(-\frac{1}{3}\right) \times 6$ — *Commutative multiplication*

$-42 \times \left(-\frac{1}{3}\right) \times 6$

14×6

84

Have student groups present their solutions to the class, describe the properties used, and explain the reasoning that supports their choices.

Closing (2 minutes)

- How do we determine the sign of expressions that include several products and quotients?
  - *We can determine the sign of the product or quotient by counting the number of negative terms. An even number of negative terms results in a positive answer. On the other hand, an odd number of negative terms results in a negative answer.*
- Name a property of operations, and describe how it is helpful when multiplying and dividing rational numbers.
  - *Answers will vary, but students should discuss the associative, commutative, and distributive properties.*

> Lesson Summary
>
> **Multiplying and dividing using the strict order of the operations in an expression is not always efficient. The properties of multiplication allow us to manipulate the expression by rearranging and regrouping factors that are easier to compute (like grouping factors 2 and 5 to get 10).**
>
> **Where division is involved, we can easily rewrite the division by a number as multiplication by its reciprocal, and then use the properties of multiplication.**
>
> **If an expression is only a product of factors, then the sign of its value is easily determined by the number of negative factors: the sign is positive if there are an even number of negative factors and negative if there an odd number of factors.**

Exit Ticket (4 minutes)

Name ______________________________ Date ______________

Lesson 16: Applying the Properties of Operations to Multiply and Divide Rational Numbers

Exit Ticket

1. Evaluate the expression below using the properties of operations.

$$18 \div \left(-\frac{2}{3}\right) \times 4 \div (-7) \times (-3) \div \left(\frac{1}{4}\right)$$

2.

a. Given the expression below, what will the sign of the product be? Justify your answer.

$$-4 \times \left(-\frac{8}{9}\right) \times 2.78 \times \left(1\frac{1}{3}\right) \times \left(-\frac{2}{5}\right) \times (-6.2) \times (-0.2873) \times \left(3\frac{1}{11}\right) \times A$$

b. Give a value for A that would result in a positive value for the expression.

c. Give a value for A that would result in a negative value for the expression.

Exit Ticket Sample Solutions

1. **Evaluate the expression below using the properties of operations.**

$$18 \div \left(-\frac{2}{3}\right) \times 4 \div (-7) \times (-3) \div \left(\frac{1}{4}\right)$$

$$18 \times \left(-\frac{3}{2}\right) \times 4 \times \left(-\frac{1}{7}\right) \times (-3) \times \left(\frac{4}{1}\right)$$

$$-27 \times 4 \times \left(-\frac{1}{7}\right) \times (-3) \times \left(\frac{4}{1}\right)$$

$$-\frac{1296}{7}$$

Answer: $-185\frac{1}{7}$ *or* $-185.\overline{142857}$

2.

a. **Given the expression below, what will the sign of the product be? Justify your answer.**

$$-4 \times \left(-\frac{8}{9}\right) \times 2.78 \times \left(1\frac{1}{3}\right) \times \left(-\frac{2}{5}\right) \times (-6.2) \times (-0.2873) \times \left(3\frac{1}{11}\right) \times A$$

There are five negative values in the expression. Because the product of two numbers with the same sign yield a positive product, pairs of negative factors have positive products. Given an odd number of negative factors, all but one can be paired into positive products. The remaining negative factor causes the product of the terms without A to be a negative value. If the value of A is negative, then the pair of negative factors forms a positive product. If the value of A is positive, the product of the two factors with opposite signs yields a negative product.

b. **Give a value for A that would result in a positive value for the expression.**

Answers will vary, but the answer must be negative. -2

c. **Give a value for A that would result in a negative value for the expression.**

Answers will vary, but the answer must be positive. 3.6

Problem Set Sample Solutions

1. **Evaluate the expression** $-2.2 \times (-2) \div \left(-\frac{1}{4}\right) \times 5$

a. **Using the order of operations only.**

$4.4 \div \left(-\frac{1}{4}\right) \times 5$

-17.6×5

-88

b. **Using the properties and methods used in Lesson 16.**

$-2.2 \times (-2) \times (-4) \times 5$

$-2.2 \times (-2) \times 5 \times (-4)$

$-2.2 \times (-10) \times (-4)$

$22 \times (-4)$

-88

c. **If you were asked to evaluate another expression, which method would you use, (a) or (b), and why?**

Answers will vary; however, most students should have found method (b) to be more efficient.

2. **Evaluate the expressions using the distributive property.**

a. $\left(2\frac{1}{4}\right) \times (-8)$

$2 \times (-8) + \frac{1}{4} \times (-8)$

$-16 + (-2)$

-18

b. $\frac{2}{3}(-7) + \frac{2}{3}(-5)$

$\frac{2}{3}(-7 + (-5))$

$\frac{2}{3}(-12)$

-8

3. **Mia evaluated the expression below but got an incorrect answer. Find Mia's error(s), find the correct value of the expression, and explain how Mia could have avoided her error(s).**

$0.38 \times 3 \div \left(-\frac{1}{20}\right) \times 5 \div (-8)$

$0.38 \times 5 \times \left(\frac{1}{20}\right) \times 3 \times (-8)$

$0.38 \times \left(\frac{1}{4}\right) \times 3 \times (-8)$

$0.38 \times \left(\frac{1}{4}\right) \times (-24)$

$0.38 \times (-6)$

-2.28

Mia made two mistakes in the second line; first, she dropped the negative symbol from $-\frac{1}{20}$ *when she changed division to multiplication. The correct term should be* (-20) *because dividing a number is equivalent to multiplying its multiplicative inverse (or reciprocal). Mia's second error occurred when she changed division to multiplication at the end of the expression; she changed only the operation, not the number. The term should be* $\left(-\frac{1}{8}\right)$. *The correct value of the expressions is* $14\frac{1}{4}$, *or* 14.25.

Mia could have avoided part of her error if she had determined the sign of the product first. There are two negative values being multiplied, so her answer should have been a positive value.

Number Correct: ______

Integer Division—Round 1

Directions: Determine the quotient of the integers, and write it in the column to the right.

1.	$4 \div 1$		23.	$-16 \div (-4)$	
2.	$4 \div (-1)$		24.	$16 \div (-2)$	
3.	$-4 \div (-1)$		25.	$-16 \div 4$	
4.	$-4 \div 1$		26.	$-20 \div 4$	
5.	$6 \div 2$		27.	$-20 \div (-4)$	
6.	$-6 \div (-2)$		28.	$-28 \div 4$	
7.	$-6 \div 2$		29.	$28 \div (-7)$	
8.	$6 \div -2$		30.	$-28 \div (-7)$	
9.	$8 \div (-4)$		31.	$-40 \div (-5)$	
10.	$-8 \div (-4)$		32.	$56 \div (-7)$	
11.	$-8 \div 4$		33.	$96 \div (-3)$	
12.	$8 \div 4$		34.	$-121 \div (-11)$	
13.	$9 \div (-3)$		35.	$169 \div (-13)$	
14.	$-9 \div 3$		36.	$-175 \div 25$	
15.	$-10 \div 5$		37.	$1 \div 4$	
16.	$10 \div (-2)$		38.	$-1 \div 4$	
17.	$-10 \div (-2)$		39.	$-1 \div (-4)$	
18.	$-10 \div (-5)$		40.	$-3 \div (-4)$	
19.	$-14 \div 7$		41.	$-5 \div 20$	
20.	$14 \div (-2)$		42.	$6 \div (-18)$	
21.	$-14 \div (-2)$		43.	$-24 \div 48$	
22.	$-14 \div (-7)$		44.	$-16 \div 64$	

Integer Division—Round 1 [KEY]

Directions: Determine the quotient of the integers, and write it in the column to the right.

1.	$4 \div 1$	4	23.	$-16 \div (-4)$	4
2.	$4 \div (-1)$	-4	24.	$16 \div (-2)$	-8
3.	$-4 \div (-1)$	4	25.	$-16 \div 4$	-4
4.	$-4 \div 1$	-4	26.	$-20 \div 4$	-5
5.	$6 \div 2$	3	27.	$-20 \div (-4)$	5
6.	$-6 \div (-2)$	3	28.	$-28 \div 4$	-7
7.	$-6 \div 2$	-3	29.	$28 \div (-7)$	-4
8.	$6 \div -2$	-3	30.	$-28 \div (-7)$	4
9.	$8 \div (-4)$	-2	31.	$-40 \div (-5)$	8
10.	$-8 \div (-4)$	2	32.	$56 \div (-7)$	-8
11.	$-8 \div 4$	-2	33.	$96 \div (-3)$	-32
12.	$8 \div 4$	2	34.	$-121 \div (-11)$	11
13.	$9 \div (-3)$	-3	35.	$169 \div (-13)$	-13
14.	$-9 \div 3$	-3	36.	$-175 \div 25$	-7
15.	$-10 \div 5$	-2	37.	$1 \div 4$	$\frac{1}{4}$
16.	$10 \div (-2)$	-5	38.	$-1 \div 4$	$-\frac{1}{4}$
17.	$-10 \div (-2)$	5	39.	$-1 \div (-4)$	$\frac{1}{4}$
18.	$-10 \div (-5)$	2	40.	$-3 \div (-4)$	$\frac{3}{4}$
19.	$-14 \div 7$	-2	41.	$-5 \div 20$	$-\frac{5}{20}$ or $-\frac{1}{4}$
20.	$14 \div (-2)$	-7	42.	$6 \div (-18)$	$-\frac{6}{18}$ or $-\frac{1}{3}$
21.	$-14 \div (-2)$	7	43.	$-24 \div 48$	$-\frac{1}{2}$
22.	$-14 \div (-7)$	2	44.	$-16 \div 64$	$-\frac{16}{64}$ or $-\frac{1}{4}$

Number Correct: ______

Improvement: ______

Integer Division—Round 2

Directions: Determine the quotient of the integers, and write it in the column to the right.

1.	$5 \div 1$		23.	$-18 \div (-9)$	
2.	$5 \div (-1)$		24.	$18 \div (-2)$	
3.	$-5 \div (-1)$		25.	$-18 \div 9$	
4.	$-5 \div 1$		26.	$-24 \div 4$	
5.	$6 \div 3$		27.	$-24 \div (-4)$	
6.	$-6 \div (-3)$		28.	$-24 \div 6$	
7.	$-6 \div 3$		29.	$30 \div (-6)$	
8.	$6 \div -3$		30.	$-30 \div (-5)$	
9.	$8 \div (-2)$		31.	$-48 \div (-6)$	
10.	$-8 \div (-2)$		32.	$64 \div (-4)$	
11.	$-8 \div 2$		33.	$105 \div (-7)$	
12.	$8 \div 2$		34.	$-144 \div (-12)$	
13.	$-9 \div (-3)$		35.	$196 \div (-14)$	
14.	$9 \div 3$		36.	$-225 \div 25$	
15.	$-12 \div 6$		37.	$2 \div 4$	
16.	$12 \div (-2)$		38.	$-2 \div 4$	
17.	$-12 \div (-2)$		39.	$-2 \div (-4)$	
18.	$-12 \div (-6)$		40.	$-4 \div (-8)$	
19.	$-16 \div 8$		41.	$-5 \div 40$	
20.	$16 \div (-2)$		42.	$6 \div (-42)$	
21.	$-16 \div (-2)$		43.	$-25 \div 75$	
22.	$-16 \div (-8)$		44.	$-18 \div 108$	

Integer Division—Round 2 [KEY]

Directions: Determine the quotient of the integers, and write it in the column to the right.

1.	$5 \div 1$	5	23.	$-18 \div (-9)$	2
2.	$5 \div (-1)$	-5	24.	$18 \div (-2)$	-9
3.	$-5 \div (-1)$	5	25.	$-18 \div 9$	-2
4.	$-5 \div 1$	-5	26.	$-24 \div 4$	-6
5.	$6 \div 3$	2	27.	$-24 \div (-4)$	6
6.	$-6 \div (-3)$	2	28.	$-24 \div 6$	-4
7.	$-6 \div 3$	-2	29.	$30 \div (-6)$	-5
8.	$6 \div -3$	-2	30.	$-30 \div (-5)$	6
9.	$8 \div (-2)$	-4	31.	$-48 \div (-6)$	8
10.	$-8 \div (-2)$	4	32.	$64 \div (-4)$	-16
11.	$-8 \div 2$	-4	33.	$105 \div (-7)$	-15
12.	$8 \div 2$	4	34.	$-144 \div (-12)$	12
13.	$-9 \div (-3)$	3	35.	$196 \div (-14)$	-14
14.	$9 \div 3$	3	36.	$-225 \div 25$	-9
15.	$-12 \div 6$	-2	37.	$2 \div 4$	$\frac{2}{4}$ or $\frac{1}{2}$
16.	$12 \div (-2)$	-6	38.	$-2 \div 4$	$-\frac{2}{4}$ or $-\frac{1}{2}$
17.	$-12 \div (-2)$	6	39.	$-2 \div (-4)$	$\frac{2}{4}$ or $\frac{1}{2}$
18.	$-12 \div (-6)$	2	40.	$-4 \div (-8)$	$\frac{4}{8}$ or $\frac{1}{2}$
19.	$-16 \div 8$	-2	41.	$-5 \div 40$	$-\frac{5}{40}$ or $-\frac{1}{8}$
20.	$16 \div (-2)$	-8	42.	$6 \div (-42)$	$-\frac{6}{42}$ or $-\frac{1}{7}$
21.	$-16 \div (-2)$	8	43.	$-25 \div 75$	$-\frac{25}{75}$ or $-\frac{1}{3}$
22.	$-16 \div (-8)$	2	44.	$-18 \div 108$	$-\frac{18}{108}$ or $-\frac{1}{6}$

Name ______________________________ Date ____________

1. Diamond used a number line to add. She started counting at 10, and then she counted until she was on the number -4 on the number line.

   a. If Diamond is modeling addition, what number did she add to 10? Use the number line below to model your answer.

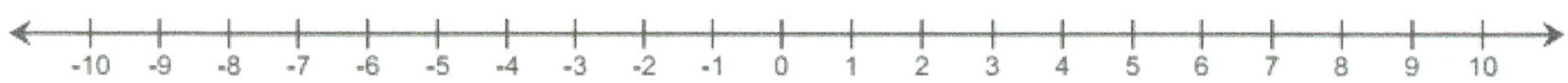

   b. Write a real-world story problem that would fit this situation.

   c. Use absolute value to express the distance between 10 and -4.

2. What value of a will make the equation a true statement? Explain how you arrived at your solution.

$$\left(-\frac{3}{4}+\frac{4}{3}\right)+a=0$$

3. Every month, Ms. Thomas pays her car loan through automatic payments (withdrawals) from her savings account. She pays the same amount on her car loan each month. At the end of the year, her savings account balance changed by $-\$2{,}931$ from payments made on her car loan.

   a. What is the change in Ms. Thomas' savings account balance each month due to her car payment?

b. Describe the total change to Ms. Thomas' savings account balance after making six monthly payments on her car loan. Model your answer using a number sentence.

4. Jesse and Miya are playing the Integer Card Game. The cards in Jesse's hand are shown below:

Jesse's Hand
$3, -5, 9, -6$

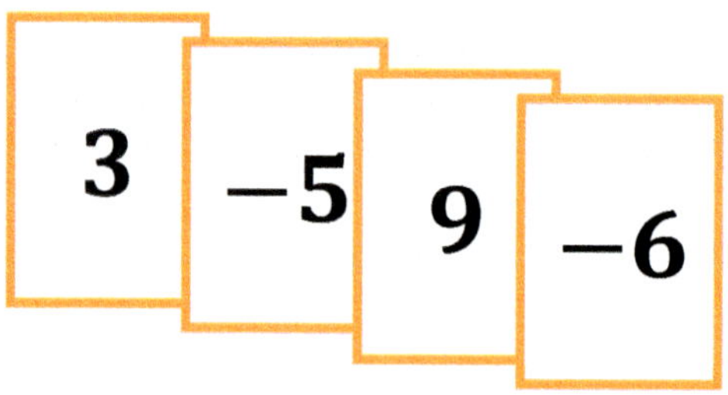

a. What is the total score of Jesse's hand? Support your answer by showing your work.

b. Jesse picks up two more cards, but they do not affect his overall point total. State the value of each of the two cards, and tell why they do not affect his overall point total.

c. Complete Jesse's new hand to make this total score equal zero. What must be the value of the **?** card? Explain how you arrived at your answer.

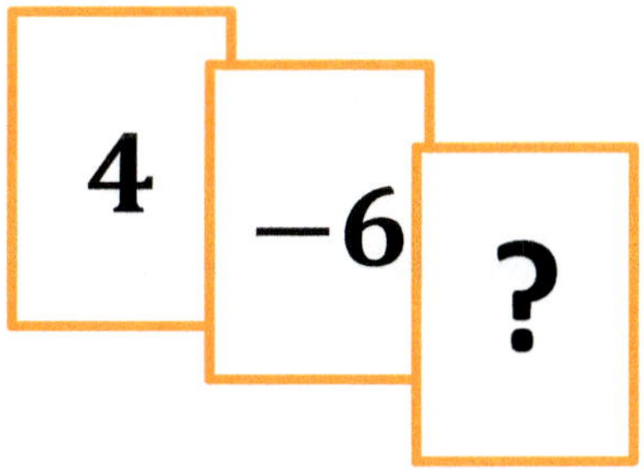

5. Michael's father bought him a 16-foot board to cut into shelves for his bedroom. Michael plans to cut the board into 11 equal size lengths for his shelves.

a. The saw blade that Michael will use to cut the board will change the length of the board by -0.125 inches for each cut. How will this affect the total length of the board?

b. After making his cuts, what will the exact length of each shelf be?

6. Bryan and Jeanette were playing the Integer Card Game like the one you played in class. They were practicing adding and subtracting integers. Jeanette had a score of -10. Bryan took away one of Jeanette's cards. He showed it to her. It was a -8. Jeanette recalculated her score to be -2, but Bryan disagreed. He said that her score should be -18 instead. Read their conversation, and answer the question below.

   "No, Jeanette, removing a negative card means the same thing as subtracting a positive. So, negative 10 *minus negative* 8 *is negative* 18*."*

   "It does not! Removing a negative card is the same as adding the same positive card. My score will go up. Negative 10 *minus negative* 8 *is negative* 2*."*

   Based on their disagreement, who, if anyone, is right? Explain.

7. The table below shows the temperature changes Monday morning in Bedford, New York over a 4-hour period after a cold front came through.

   a. If the beginning temperature was $-13°\text{F}$ at 5:00 a.m., what was the temperature at 9:00 a.m.?

Change in Temperature	
5:00 a.m. – 6:00 a.m.	$-3°\text{F}$
6:00 a.m. – 7:00 a.m.	$-2°\text{F}$
7:00 a.m. – 8:00 a.m.	$-6°\text{F}$
8:00 a.m. – 9:00 a.m.	$7°\text{F}$

   b. The same cold front hit Hartford, Connecticut the next morning. The temperature dropped by $7°\text{F}$ each hour from 5:00 a.m. to 9:00 a.m. What was the beginning temperature at 5:00 a.m. if the temperature at 9:00 a.m. was $-10°\text{F}$?

   c. In answering part (b), Josiah and Kate used different methods. Josiah said his method involved multiplication, while Kate said she did not use multiplication. Both students arrived at the correct answer. How is this possible? Explain.

A Progression Toward Mastery

Assessment Task Item		STEP 1 Missing or incorrect answer and little evidence of reasoning or application of mathematics to solve the problem.	STEP 2 Missing or incorrect answer but evidence of some reasoning or application of mathematics to solve the problem.	STEP 3 A correct answer with some evidence of reasoning or application of mathematics to solve the problem, OR an incorrect answer with substantial evidence of solid reasoning or application of mathematics to solve the problem.	STEP 4 A correct answer supported by substantial evidence of solid reasoning or application of mathematics to solve the problem.
1	a 7.NS.A.1b	Student answer is incorrect. Student attempts to model answer using vector addition but has more than 2 missing parts. OR Student answer is incorrect and does not use the number line.	Student answer is incorrect. Student attempts to model the answer using vector addition but has 1–2 missing parts (i.e., only one vector displayed, wrong direction or incorrect starting or ending point).	Student answer is incorrect due to a minor mistake modeling the answer with vector addition. OR Student shows vector addition modeled correctly, but student records the incorrect answer. For example, student miscounts and ends the second vector at -5 or -3.	Student correctly answers -14 and models the answer using vector addition on the number line with both arrows showing correct direction, starting point, and ending point.
	b 7.NS.A.1b	Student answer is incorrect. No story problem is created. OR Student's story problem is not real world and has significant errors such as little to no appropriate vocabulary in context to reflect addition of positive and negative numbers.	Student incorrectly interprets $10 + (-14) = -4$ and begins to create a real-world story problem; however, the story is incomplete. OR Student's story problem shows some evidence of a correct interpretation of $10 + (-14) = -4$, but it is not cohesive.	Student correctly interprets $10 + (-14) = -4$ by creating a relevant real-world story problem but makes an incorrect statement or incorrect use of vocabulary. For example, student describes -14 as a deposit of $\$14$. OR Student creates a relevant real-world story problem based on an incorrect sum.	Student correctly interprets $10 + (-14) = -4$ by creating a relevant real-world story problem and uses appropriate and accurate vocabulary in context to reflect addition of positive and negative numbers.

	c 7.NS.A.1c	Student answer is incorrect with little or no evidence of reasoning. Student does not use absolute value notation.	Student answer is incomplete but shows some evidence of reasoning. For instance, student represents the distance as $\lvert 10-(-4)\rvert$, but no further correct work is shown.	Student correctly answers $\lvert 10-(-4)\rvert = 14$ but does not include steps that show subtraction as addition of the additive inverse in order to derive the correct answer. For example, student shows $\lvert 14\rvert = 14$ only.	Student correctly answers $\lvert 10-(-4)\rvert = 14$ and shows evidence of reasoning by including steps that show subtraction as addition of the additive inverse in order to arrive at the correct answer.
2	7.NS.A.1b 7.NS.A.1c	Student answer is incorrect or missing. Student shows little or no evidence of understanding how to find the sum of the fractions and the opposite of the sum.	Student answer is incorrect, but explanation and/or work shows some evidence of reasoning.	Student answer is incorrect, but work and/or explanation shows solid evidence of reasoning. For example, student includes a negative on the sum $\left(-\frac{7}{12}\right)$ and gives a positive additive inverse $\left(\frac{7}{12}\right)$ as a final answer. OR Student arrives at the correct sum of $\frac{7}{12}$, but the explanation is incomplete.	Student correctly answers $a = -\frac{7}{12}$ as the additive inverse and has no errors in the steps taken to arrive at the answer.
3	a 7.NS.A.2a 7.NS.A.2b	Student answer is incorrect or missing. Student shows little or no evidence of understanding the long division process.	Student answer is incorrect, but student begins the process of dividing 2,931 by 12.	Student correctly states $-\$244.25$ as the monthly change in account balance, but the work is incomplete. OR Student incorrectly answers due to a minor calculation error in the long division process.	Student correctly states $-\$244.25$ as the monthly change in account balance and provides the correct long division work to show that 2,931 divided by 12 equals 244.25.

	b 7.NS.A.2a 7.NS.A.2b	Student answer is incorrect or missing. Student shows little or no work and does not provide a number sentence.	Student answer is incorrect, but student demonstrates an understanding of the task involved by multiplying 244.25×6 or uses another method to indicate the multiplication process.	Student correctly describes the six-month change in the account balance as $-\$1{,}465.50$ but fails to provide a correct number sentence. OR Student uses a correct method but incorrectly describes the six-month change due to a minor calculation error, which is reflected in the number sentence.	Student correctly describes the six-month change in the account balance as $-\$1{,}465.50$ and provides a correct number sentence, such as $6(-244.25) = -1{,}465.50$.
4	**a** 7.NS.A.1a 7.NS.A.1b	Student answer is incorrect or missing. Student shows little or no evidence of understanding how to add integers.	Student uses a correct representation to find the sum of $3 + (-5) + 9 + (-6)$ but has several errors in the process.	Student correctly states a score of 1 but does not provide enough work to support the answer.	Student correctly states a score of 1 and creates an expression to represent the sum of all four card values, correctly adding the integers.
	b 7.NS.A.1a 7.NS.A.1b	Student does not state the values of 2 cards whose sum is zero and is unable to provide a correct written explanation.	Student explains that the 2 cards must total zero but does not correctly state 2 integers whose sum is zero.	Student correctly states the values of two opposite numbers, such as -2 and 2 but does not provide a complete written explanation to tell why they do not affect the overall point total.	Student correctly states the values of two opposite numbers, such as -2 and 2, and explains that because they are opposites, their sum is zero, so the overall point total will not change.
	c 7.NS.A.1a 7.NS.A.1b	Student answer is incorrect or missing. Student work shows little or no evidence of understanding of adding 4 and -6 and then finding the opposite of their sum.	Student completes the first step by adding 4 and -6 to arrive at -2 but does not complete any further correct steps. Student's explanation is incomplete.	Student correctly answers 2 but does not justify the answer through a written response. OR Student incorrectly answers -2 but correctly explains the process of finding the sum of -6 and 4 and then finding the opposite of their sum.	Student correctly answers 2, and the explanation of how he or she arrived at 2 indicates a sound understanding of finding the sum of the 4 and -6 and then finding the opposite of the sum.

5	a 7.NS.A.2 7.NS.A.3	Student answer is incorrect or missing. Student work shows little or no evidence of understanding of the process involved.	Student answer is incomplete. The student states the board would be shortened but does not determine the amount of change in the board's length.	Student uses a sound process and shows solid evidence of reasoning. Student knows that the board length would be shortened, but the amount stated is incorrect due to a minor calculation error. OR Student multiplies -0.125×10 to get -1.25 but does not provide a written explanation to interpret -1.25 in the context of the situation.	Student correctly answers that the length of the board will be shortened by 1.25 inches and shows the correct calculations and/or diagram to support the answer.
	b 7.NS.A.2 7.NS.A.3	Student answer is incorrect or missing. Student work shows little or no evidence of correct reasoning.	Student answer is incorrect. Student work shows some understanding of the steps involved, and there is evidence of the division process.	Student incorrectly answers the exact length of each shelf to be $1.\overline{45}$ feet and shows work for dividing 16 by 11 to arrive at the answer. OR Student follows the correct process of subtracting 1.25 inches from 192 inches and then dividing 190.75 by 11 but arrives at an incorrect answer due to a computational error in the long division process.	Student correctly answers the exact length of each shelf to be $17.34\overline{09}$ inches, and the correct steps and long division work are shown to support the answer.
6	7.NS.A.1c	Student answer is incorrect or missing. Student work shows little or no evidence of correct reasoning.	Student answer is incorrect. Student uses evidence of some reasoning to justify answer but does not know the rule for subtracting negative numbers and cannot apply it in context.	Student correctly answers *Jeanette*, uses evidence of reasoning by knowing the rule for subtracting negative numbers, and provides some justification of the correct answer by applying this rule in context. For example, student may only paraphrase Jeanette's statement with no further explanation of his or her own.	Student correctly answers *Jeanette*, uses solid evidence of reasoning by knowing the rule for subtracting negative numbers, and provides substantial justification of the correct answer by applying this rule in context with further explanation or a mathematical model.

7	a 7.NS.A.1d	Student answers incorrectly with little or no evidence of understanding of how to add integers.	Student does not arrive at the correct answer. Student work indicates some degree of understanding, as at least one pair of integers is correctly added.	Student correctly answers $-17°\text{F}$ but does not show sufficient work to support the answer.	Student correctly answers $-17°\text{F}$ and has no errors in the steps taken to arrive at the answer.
	b 7.NS.A.3	Student answers incorrectly with little or no evidence of understanding how to work backward to find the beginning temperature.	Student answers incorrectly but is able to set up a correct visual model or numerical expression, such as $-10-(-7)(4)$, to represent the situation.	Student correctly answers $18°\text{F}$ but student's work is incomplete. OR Student answers incorrectly due to a calculation error but uses the correct process.	Student correctly answers $18°\text{F}$ and has no errors in the steps taken to arrive at the answer.
	c 7.NS.A.1b 7.NS.A.2a	Student provides no explanation. OR Student states that the situation is not possible.	Student answers only part of the question. For instance, student explains how Josiah used multiplication to arrive at a correct answer but is unable to explain how Kate used a different operation to arrive at the same answer.	Student indicates an understanding of multiplication as repeated addition, but the explanation in student's written response is not complete.	Student responds by indicating that a drop of 7 degrees four times can be represented by multiplication or repeated addition: $(7)+(7)+(7)+(7)$. OR Student responds by indicating that a drop of 7 degrees four times can be represented by multiplication or repeated subtraction: $-(-7)-(-7)-(-7)-(-7)$.

Name ______________________________________ Date ________________

1. Diamond used a number line to add. She started counting at 10, and then she counted until she was on the number -4 on the number line.

   a. If Diamond is modeling addition, what number did she add to 10? Use the number line below to model your answer.

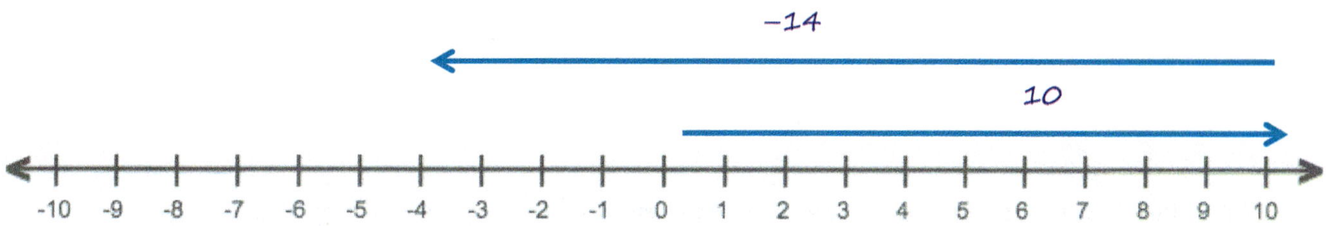

   $10 + -14 = -4$

   Diamond added -14 to 10.

   b. Write a real-world story problem that would fit this situation.

   Diamond had \$10 and put it in the bank. She forgot about the monthly bank fee of \$14.

   Now her account has a balance of -\$4.

   c. Use absolute value to express the distance between 10 and -4.

   $|10 - (-4)|$

   $|10 + 4|$

   $|14|$

   14

   The distance between 10 and -4 is 14.

2. What value of a will make the equation a true statement? Explain how you arrived at your solution.

$$\left(-\frac{3}{4}+\frac{4}{3}\right)+a=0$$

$$\left(\frac{-3}{4}+\frac{4}{3}\right)$$

$$\left(\frac{-9}{12}+\frac{16}{12}\right)$$

$$\frac{7}{12} \qquad\qquad \frac{7}{12}+a=0$$

$$a=-\frac{7}{12}$$

The value of a has to be $-\frac{7}{12}$ because that's the additive inverse of $\frac{7}{12}$.

3. Every month, Ms. Thomas pays her car loan through automatic payments (withdrawals) from her savings account. She pays the same amount on her car loan each month. At the end of the year, her savings account balance changed by $-\$2,931$ from payments made on her car loan.

   a. What is the change in Ms. Thomas' savings account balance each month due to her car payment?

```
       0 2 4 4 . 2 5
    ______________
1 2 | 2 9 3 1 . 0 0
      0
      ---
      2 9
    - 2 4
      ---
        5 3
        4 8
        ---
          5 1
          4 8
          ---
            3 0
            2 4
            ---
              6 0
              6 0
              ---
                0
```

Her monthly payment is \$244.25, so her account balance changes each month by $-\$244.25$ when her payment is made.

b. Describe the total change to Ms. Thomas' savings account balance after making six monthly payments on her car loan. Model your answer using a number sentence.

$$\begin{array}{r} {}^{2}2\;{}^{2}4\;{}^{1}4\,.\,{}^{3}2\;5 \\ \times \qquad\qquad 6 \\ \hline 1\;4\;6\;5\,.\,5\;0 \end{array}$$

$6 \times (-244.25) = -1{,}465.50$

Ms. Thomas' car loan changed her savings account balance by −$1,465.50 after 6 monthly payments.

4. Jesse and Miya are playing the Integer Card Game. The cards in Jesse's hand are shown below:

Jesse's Hand
$3, -5, 9, -6$

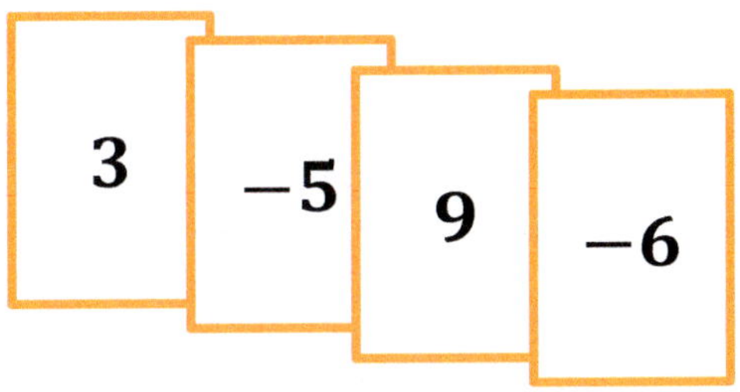

a. What is the total score of Jesse's hand? Support your answer by showing your work.

$3 + (-5) + 9 + (-6)$
$(-2) + 3$
1
Jesse's score is 1.

b. Jesse picks up two more cards, but they do not affect his overall point total. State the value of each of the two cards, and tell why they do not affect his overall point total.

The values of the two cards must be opposites, such as −2 and 2, because opposites combine to get 0. 0 will not change the score in his hand.

c. Complete Jesse's new hand to make this total score equal zero. What must be the value of the **?** card? Explain how you arrived at your answer.

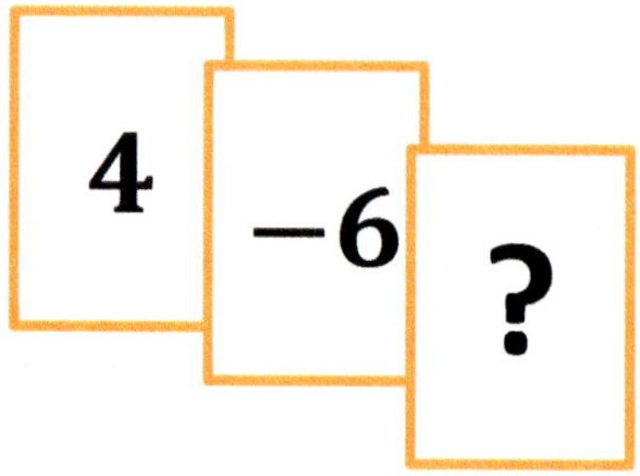

$4 + (-6)$

-2

$-2 + a = 0$

$a = 2$

The two given cards total −2. To get a sum of zero, I have to combine −2 with its opposite 2 because additive inverses (opposites) combine to get 0.

5. Michael's father bought him a 16-foot board to cut into shelves for his bedroom. Michael plans to cut the board into 11 equal size lengths for his shelves.

   a. The saw blade that Michael will use to cut the board will change the length of the board by -0.125 inches for each cut. How will this affect the total length of the board?

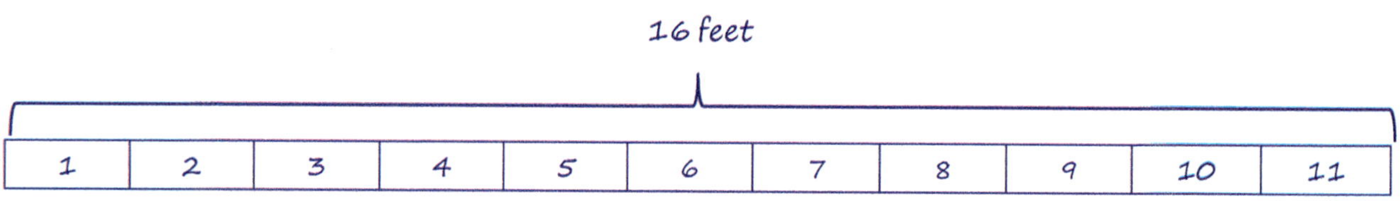

The board will be cut in 10 places.

$$\begin{array}{r} 0.125 \\ \times \quad 10 \\ \hline 1.250 \end{array}$$

$-0.125 \times 10 = -1.25$

The 10 cuts take away 1.25 inches of the total length of the board. The usable length of the board is 1.25 inches shorter than 16 feet.

b. After making his cuts, what will the exact length of each shelf be?

The board begins at 192 inches long.

$192 - 1.25 = 190.75$. The length of the board that can be used with the blade widths removed is 190.75 inches.

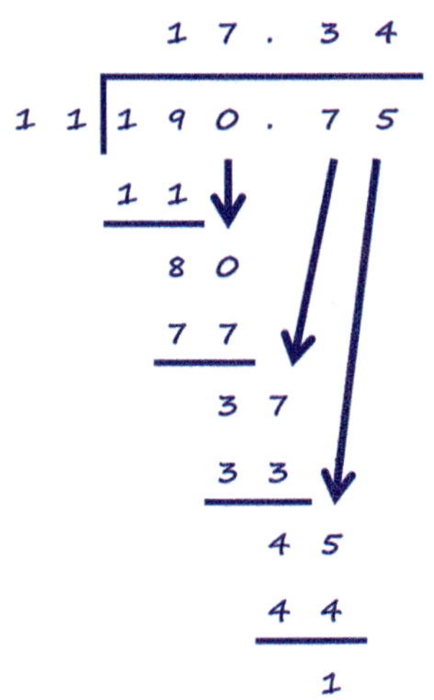

Continue the long division, and there will be a repeating remainder of 1. Therefore, the lengths of the shelves should be exactly $17.34\overline{09}$ inches.

6. Bryan and Jeanette were playing the Integer Card Game like the one you played in class. They were practicing adding and subtracting integers. Jeanette had a score of -10. Bryan took away one of Jeanette's cards. He showed it to her. It was a -8. Jeanette recalculated her score to be -2, but Bryan disagreed. He said that her score should be -18 instead. Read their conversation and answer the question below.

"No Jeanette, removing a negative card means the same thing as subtracting a positive. So, negative 10 *minus negative* 8 *is negative* 18.*"*

"It does not! Removing a negative card is the same as adding the same positive card. My score will go up. Negative 10 *minus negative* 8 *is negative* 2.*"*

Based on their disagreement, who, if anyone, is right? Explain.

Jeanette is correct that removing a negative is the same as adding the same positive card. Having a negative card in your hand decreases your score. If you remove that negative card, your score is no longer decreased by the card, so your score goes up.

7. The table below shows the temperature changes Monday morning in Bedford, New York over a 4-hour period after a cold front came through.

   a. If the beginning temperature was $-13°\text{F}$ at 5:00 a.m., what was the temperature at 9:00 a.m.?

Change in Temperature	
5:00 a.m. – 6:00 a.m.	$-3°\text{F}$
6:00 a.m. – 7:00 a.m.	$-2°\text{F}$
7:00 a.m. – 8:00 a.m.	$-6°\text{F}$
8:00 a.m. – 9:00 a.m.	$7°\text{F}$

$$-13 + (-3) + (-2) + (-6) + 7$$

$$(-16) + (-8) + 7$$

$$-24 + 7$$

$$-17°\text{F}$$

The temperature at 9:00 a.m. was $-17°\text{F}$.

   b. The same cold front hit Hartford, Connecticut the next morning. The temperature dropped by $7°\text{F}$ each hour from 5:00 a.m. to 9:00 a.m. What was the beginning temperature at 5:00 a.m. if the temperature at 9:00 a.m. was $-10°\text{F}$?

$$-10 + 4(7)$$

$$-10 + 28$$

$$18°\text{F}$$

The beginning temperature at 5:00 a.m. was $18°\text{F}$.

   c. In answering part (b), Josiah and Kate used different methods. Josiah said his method involved multiplication, while Kate said she did not use multiplication. Both students arrived at the correct answer. How is this possible? Explain.

The temperature change was the same for each hour, so Josiah multiplied the 7-degree drop by 4 hours. Kate added the 7-degree drop 4 times. Kate used repeated addition, which is the same as multiplication.

Topic C

Applying Operations with Rational Numbers to Expressions and Equations

7.NS.A.3, 7.EE.A.2, 7.EE.B.4a

Focus Standards:	7.NS.A.3	Solve real-world and mathematical problems involving the four operations with rational numbers.
	7.EE.A.2	Understand that rewriting an expression in different forms in a problem context can shed light on the problem and how the quantities in it are related. *For example,* $a + 0.05a = 1.05a$, *means that "increase by* 5%*" is the same as "multiply by* 1.05*".*
	7.EE.B.4	Use variables to represent quantities in a real-world or mathematical problem, and construct simple equations and inequalities to solve problems by reasoning about quantities.
		a. Solve word problems leading to equations of the form $px + q = r$ and $p(x + q) = r$, where p, q, and r are specific rational numbers. Solve equations of these forms fluently. Compare an algebraic solution to an arithmetic solution, identifying the sequence of the operations used in each approach. *For example, the perimeter of a rectangle is* 54 cm. *Its length is* 6 cm. *What is its width?*
Instructional Days:	7	
Lesson 17:	Comparing Tape Diagram Solutions to Algebraic Solutions (P)[1]	
Lessons 18–19:	Writing, Evaluating, and Finding Equivalent Expressions with Rational Numbers (P, P)	
Lesson 20:	Investments—Performing Operations with Rational Numbers (M)	
Lesson 21:	If-Then Moves with Integer Number Cards (E)	
Lessons 22–23:	Solving Equations Using Algebra (P, P)	

[1]Lesson Structure Key: **P**-Problem Set Lesson, **M**-Modeling Cycle Lesson, **E**-Exploration Lesson, **S**-Socratic Lesson

Students use algebra and rational numbers in Topic C to problem-solve, building upon their foundational work with rational numbers and expressions and equations in Grade 6 (**6.NS.C.5**, **6.EE.A.2**, **6.EE.A.3**, **6.EE.A.4**, **6.EE.B.5**, **6.EE.B.6**, **6.EE.B.7**). Topic C begins in Lesson 17 with students finding solutions to word problems by working backward and using tape diagrams to model the algebraic steps they use to arrive at the solution. In Lessons 18 and 19, students create and evaluate equivalent forms of expressions involving rational numbers to see structure, reveal characteristics, and make connections to context (**7.EE.A.2**). Lesson 20 is a modeling lesson in which students are presented with a scenario related to an investment account's activity over the course of several years. Students interpret the information and develop a strategy to find the actual changes to the account balance each year. In Lesson 21, students return to the Integer Game that they played in earlier lessons to better understand *if-then* statements. They relate making the same changes to two equal card-hand totals to making equivalent changes to each side of a true number sentence. Therefore, they show, for instance: *If* $a = b$, *then* $a - c = b - c$. Topic C concludes with a two-day lesson. In Lessons 22 and 23, students work toward fluently solving word problems through the use of equations (**7.EE.B.4a**). Using algebra to deconstruct and solve contextual problems continues as the focus in Module 3.

Lesson 17: Comparing Tape Diagram Solutions to Algebraic Solutions

Student Outcomes

- Students use tape diagrams to solve equations of the form $px + q = r$ and $p(x + q) = r$ (where p, q, and r are *small positive* integers) and identify the sequence of operations used to find the solution.
- Students translate word problems to write and solve algebraic equations using tape diagrams to model the steps they record algebraically.

Lesson Notes

In Lesson 17, students relate their algebraic steps in solving an equation to the steps they take to arrive at the solution arithmetically. They refer back to the use of tape diagrams to conceptually understand the algebraic steps taken to solve an equation. It is not until Lesson 21 that students use the Properties of Equality to formally justify performing the same operation to both sides of the equation. Teachers need to have scenarios ready for each group. There should be enough copies of each scenario for each person in the group so they can bring this information with them to the next group.

Classwork

Opening Exercise (Optional, 5 minutes)

Students solve two-step equations for the first time throughout the Exploratory Challenge. The Opening Exercise is an optional way to provide some guidance to students before they begin working in their groups or if students struggle during the Exploratory Challenge. It is important for students to have the opportunity to apply their previous knowledge of equations without too much guidance from the teacher.

Opening Exercise

For his birthday, Zack and three of his friends went to a movie. They each got a ticket for $\$8.00$ and the same snack from the concession stand. If Zack's mom paid $\$48$ for the group's tickets and snacks, how much did each snack cost?

The equation $4(s + 8) = 48$ represents the situation when s represents the cost, in dollars, of one snack.

- Use your knowledge of equations to determine the cost of each snack.

Provide students with time to try different methods to answer the question. As students work, the following questions may be helpful in guiding them through their work.

- What information do you know?
  - *I know the total cost, the price of the ticket, and how many friends attended the movie.*
- What are you trying to figure out?
  - *I am trying to determine the cost of each snack.*

- Can you manipulate the information you know to answer the question?
  - *Yes. I can figure out the total amount spent on movie tickets and use this information to determine the total amount spent on snacks.*

If students are able to answer the question without a drawing or equation, encourage students to use their methods of solving to write an equation and create a tape diagram.

$$\begin{aligned} 4(s+8) &= 48 \\ \frac{1}{4}(4(s+8)) &= \frac{1}{4}(48) \\ s+8 &= 12 \\ s+8-8 &= 12-8 \\ s+0 &= 4 \\ s &= 4 \end{aligned}$$

OR

$$\begin{aligned} 4(s+8) &= 48 \\ 4s+32 &= 48 \\ 4s+32-32 &= 48-32 \\ 4s+0 &= 16 \\ 4s &= 16 \\ \frac{1}{4}(4s) &= \frac{1}{4}(16) \\ 1s &= 4 \\ s &= 4 \end{aligned}$$

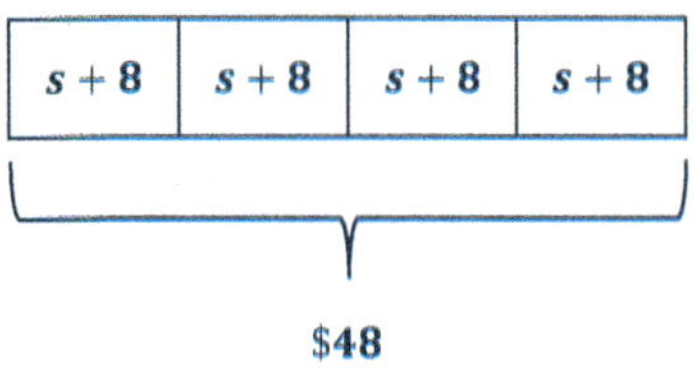

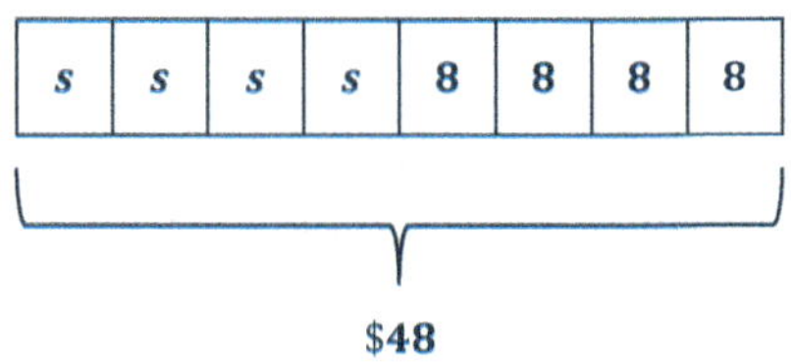

Exploratory Challenge (30 minutes): Expenses on Your Family Vacation

MP.1 & MP.4

Divide students into seven groups. Each group is responsible for one of the seven specific expense scenarios. Provide each student in the group with a copy of the group's assigned scenario, which is found at the end of this lesson. In these groups, students write algebraic equations and solve by modeling the problem using a tape diagram. Then have groups collaborate to arrive at the sequence of operations used to find the solution. Lastly, challenge students to show an algebraic solution to the same problem.

> *Scaffolding:*
>
> Review how to set up a tape diagram when given the parts and total.

Once groups work on an individual scenario, mix up the groups so that each group now has seven students (i.e., one student to represent each of the seven expenses). Within each group, students present their specific scenario to the other members of the group: the solution and model used to find the solution, the sequence of operations used, and a possible algebraic solution. After all scenarios have been shared, and each student completes the summary sheet, have students answer the questions regarding total cost for several different combinations.

Exploratory Challenge: Expenses on Your Family Vacation

John and Ag are summarizing some of the expenses of their family vacation for themselves and their three children, Louie, Missy, and Bonnie. Write an algebraic equation, create a model to determine how much each item will cost using all of the given information, and answer the questions that follow.

Expenses:

Car and insurance fees: $\$400$	**Airfare and insurance fees:** $\$875$	**Motel and tax:** $\$400$
Baseball game and hats: $\$103.83$	**Movies for one day:** $\$75$	**Soda and pizza:** $\$37.95$
	Sandals and T-shirts: $\$120$	

Your Group's Scenario Solution:

Scenario 1

During one rainy day on the vacation, the entire family decided to go watch a matinee movie in the morning and a drive-in movie in the evening. The price for a matinee movie in the morning is different than the cost of a drive-in movie in the evening. The tickets for the matinee morning movie cost $\$6$ each. How much did each person spend that day on movie tickets if the ticket cost for each family member was the same? What was the cost for a ticket for the drive-in movie in the evening?

Algebraic Equation & Solution

Morning matinee movie: $\$6$ *each*

Evening drive-in movie: e *each*

$$5(e+6)=75$$
$$5e+30=75$$
$$5e+30-30=75-30$$
$$5e+0=45$$
$$\left(\frac{1}{5}\right)5e=45\left(\frac{1}{5}\right)$$
$$1e=9$$
$$e=9$$

OR

$$5(e+6)=75$$
$$\left(\frac{1}{5}\right)5(e+6)=75\left(\frac{1}{5}\right)$$
$$e+6=15$$
$$e+6-6=15-6$$
$$e=9$$

Tape Diagram

Total: 75

$e+6$	$e+6$	$e+6$	$e+6$	$e+6$
John	*Ag*	*Louie*	*Missy*	*Bonnie*

$5e$

$5(6)=30$

$75-30=45$

$45\div 5=9$

The total each person spent on movies in one day was $\$15$*. The evening drive-in movie costs* $\$9$ *each.*

Scenario 2

For dinner one night, the family went to the local pizza parlor. The cost of a soda was $\$3$. If each member of the family had a soda and one slice of pizza, how much did one slice of pizza cost?

Algebraic Equation & Solution

One Soda: $\$3$

Slice of Pizza: p *dollars*

$$5(p+3)=37.95$$
$$5p+15=37.95$$
$$5p+15-15=37.95-15$$
$$5p+0=22.95$$
$$\left(\frac{1}{5}\right)5p=22.95\left(\frac{1}{5}\right)$$
$$1p=4.59$$
$$p=4.59$$

OR

$$5(p+3)=37.95$$
$$\left(\frac{1}{5}\right)5(p+3)=(37.95)\left(\frac{1}{5}\right)$$
$$p+3=7.59$$
$$p+3-3=7.59-3$$
$$p=4.59$$

Tape Diagram

Total: 37.95

$p+3$	$p+3$	$p+3$	$p+3$	$p+3$
John	*Ag*	*Louie*	*Missy*	*Bonnie*

$5(3)=15$

$5p$

$37.95-15=22.95$

$22.95\div 5=4.59$

One slice of pizza costs $\$4.59$*.*

Scenario 3

One night, John, Louie, and Bonnie went to see the local baseball team play a game. They each bought a game ticket and a hat that cost $\$10$. How much was each ticket to enter the ballpark?

Algebraic Equation & Solution

Ticket: t dollars

Hat: $\$10$

$$3(t+10)=103.83$$
$$3t+30=103.83$$
$$3t+30-30=103.83-30$$
$$3t+0=73.83$$
$$\left(\frac{1}{3}\right)3t=73.83\left(\frac{1}{3}\right)$$
$$1t=24.61$$
$$t=24.61$$

OR

$$3(t+10)=103.83$$
$$\left(\frac{1}{3}\right)3(t+10)=(103.83)\left(\frac{1}{3}\right)$$
$$t+10=34.61$$
$$t+10-10=34.61-10$$
$$t=24.61$$

Tape Diagram

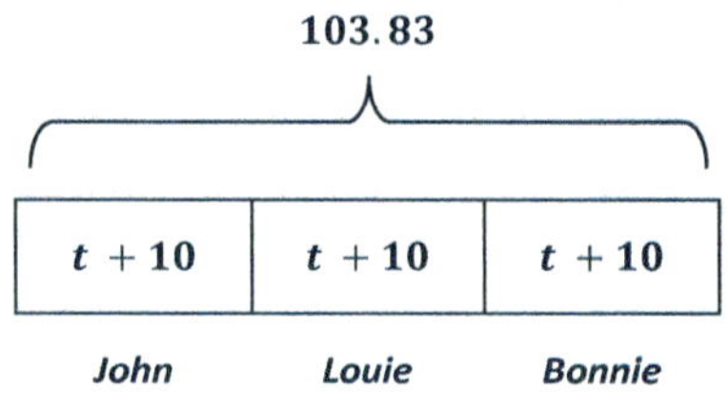

$3(10)=30$ $\quad 103.83-30=73.83$

$73.83\div 3=24.61$

One ticket costs $\$24.61$.

Scenario 4

While John, Louie, and Bonnie went to see the baseball game, Ag and Missy went shopping. They bought a T-shirt for each member of the family and bought two pairs of sandals that cost $\$10$ a pair. How much was each T-shirt?

Algebraic Equation & Solution

T-Shirt: t dollars

Sandals: $2\times \$10=\$20$

$$5t+20=120$$
$$5t+20-20=120-20$$
$$5t+0=100$$
$$\left(\frac{1}{5}\right)5t=100\left(\frac{1}{5}\right)$$
$$1t=20$$
$$t=20$$

Tape Diagram

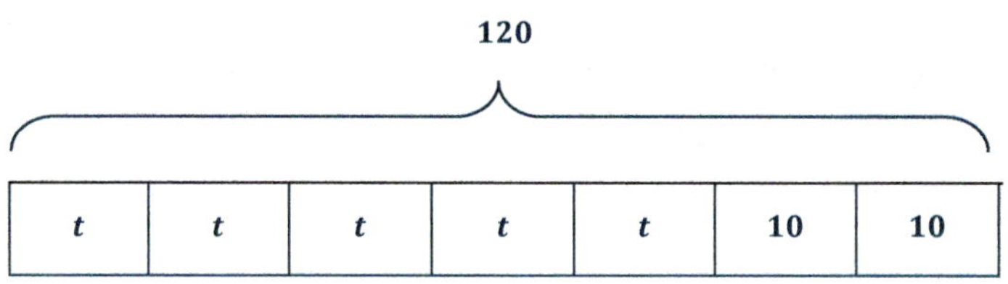

$2(10)=20$ $\quad 120-20=100$ $\quad 100\div 5=20$

One T-shirt costs $\$20$.

Scenario 5

The family flew in an airplane to their vacation destination. Each person had to have his own ticket for the plane and also pay $\$25$ in insurance fees per person. What was the cost of one ticket?

Algebraic Equation & Solution

One ticket: t dollars

Insurance: $\$25$ per person

$$5(t + 25) = 875$$
$$5t + 125 - 125 = 875 - 125$$
$$5t + 0 = 750$$
$$5t = 750\left(\frac{1}{5}\right)$$
$$t = 150$$

OR

$$5(t + 25) = 875$$
$$\left(\frac{1}{5}\right)5(t+25) = \left(\frac{1}{5}\right)(875)$$
$$t + 25 = 175$$
$$t + 25 - 25 = 175 - 25$$
$$t = 150$$

Tape Diagram

875

$t+25$	$t+25$	$t+25$	$t+25$	$t+25$
John	*Ag*	*Missy*	*Louie*	*Bonnie*

$5(25) = 125$

$875 - 125 = 750$

$750 \div 5 = 150$

One ticket costs $\$150$.

Scenario 6

While on vacation, the family rented a car to get them to all the places they wanted to see for five days. The car costs a certain amount each day, plus a one-time insurance fee of $\$50$. How much was the daily cost of the car (not including the insurance fees)?

Algebraic Equation & Solution

Daily fee: d dollars

Insurance fee: $\$50$

$$5d + 50 = 400$$
$$5d + 50 - 50 = 400 - 50$$
$$5d + 0 = 350$$
$$\left(\frac{1}{5}\right)5d = 350\left(\frac{1}{5}\right)$$
$$1d = 70$$
$$d = 70$$

Tape Diagram

400

d	d	d	d	d	50
Day 1	*Day 2*	*Day 3*	*Day 4*	*Day 5*	*Insurance*

$1(50) = 50$

$5d$

$400 - 50 = 350$

$350 \div 5 = 70$

One day costs $\$70$.

Scenario 7

The family decided to stay in a motel for four nights. The motel charges a nightly fee plus $\$60$ in state taxes. What is the nightly charge with no taxes included?

Algebraic Equation & Solution

Nightly charge: n dollars

Taxes: $\$60$

$$4n + 60 = 400$$
$$4n + 60 - 60 = 400 - 60$$
$$4n + 0 = 340$$
$$\left(\frac{1}{4}\right)4n = 340\left(\frac{1}{4}\right)$$
$$1n = 85$$
$$n = 85$$

Tape Diagram

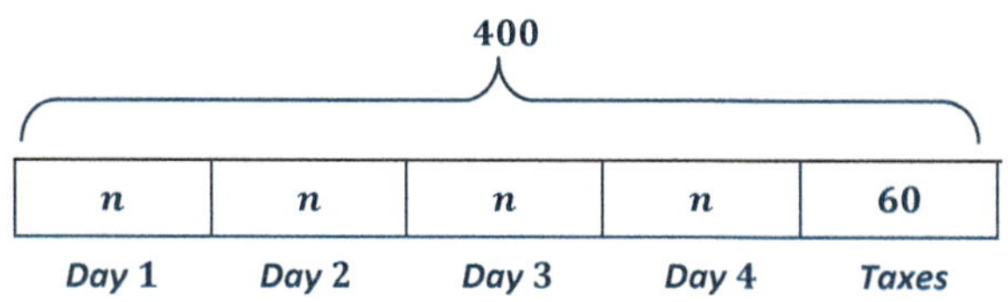

$4n \quad 400 - 60 = 340 \quad 340 \div 4 = 85$

One night costs $\$85$.

Once students have completed their group scenario to determine the cost of the item, and the groups have been mixed so students have seen the problems and solutions to each expense, have them complete the summary chart and answer the questions that follow.

After collaborating with all of the groups, summarize the findings in the table below.

Cost of Evening Movie	$\$9$
Cost of 1 Slice of Pizza	$\$4.59$
Cost of the Admission Ticket to the Baseball Game	$\$24.61$
Cost of 1 T-Shirt	$\$20$
Cost of 1 Airplane Ticket	$\$150$
Daily Cost for Car Rental	$\$70$
Nightly Charge for Motel	$\$85$

Using the results, determine the cost of the following:

1. **A slice of pizza, 1 plane ticket, 2 nights in the motel, and 1 evening movie.**

   $4.59 + 150 + 2(85) + 9 = 333.59$

2. **One T-shirt, 1 ticket to the baseball game, and 1 day of the rental car.**

   $20 + 24.61 + 70 = 114.61$

Discussion/Lesson Questions for Algebraic Approach

Scaffolding:

Review from Grade 6 solving 1-step and 2-step equations algebraically, as well as the application of the distributive property.

MP.2

The importance of undoing addition and multiplication to get 0 and 1 (i.e., using the additive inverse undoes addition to get 0, and the multiplicative inverse undoes multiplication by a nonzero number to get 1) should be stressed.

- When solving an equation with parentheses, order of operations must be followed. What property can be used to eliminate parentheses; for example, $3(a+b) = 3a + 3b$?
  - *To eliminate parentheses, the distributive property must be applied.*
- Another approach to solving the problem is to eliminate the coefficient first. How would one go about eliminating the coefficient?
  - *To eliminate the coefficient, you can multiply both sides by the reciprocal of the coefficient or divide both sides by the coefficient.*
- How do we *undo* multiplication?
  - *We multiply by the reciprocal of the coefficient of the variable or divide both sides of the equation by the coefficient.*
- What is the result when *undoing* multiplication in any problem?
  - *When undoing multiplication, the result will always be* 1*, which is the multiplicative identity.*
- What mathematical property is being applied when *undoing* multiplication?
  - *Multiplicative inverse*
- What approach must be taken when solving for a variable in an equation and *undoing* addition is required?
  - *To undo addition, you need to subtract the constant.*
- How can this approach be shown with a tape diagram?

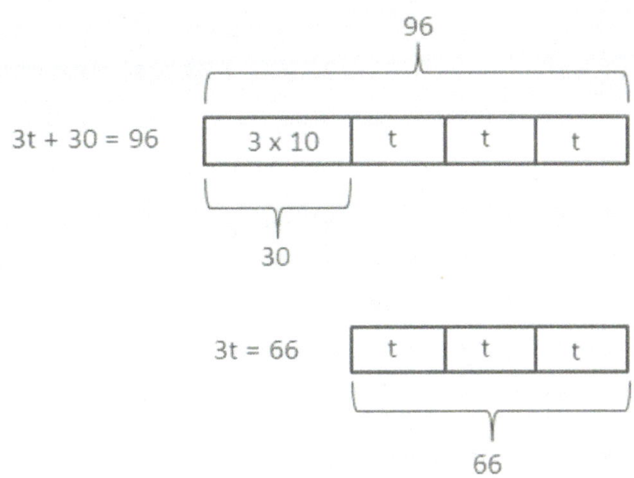

- What is the result when *undoing* addition in any problem?
  - *The result will always be* 0*, which is the additive identity.*
- What mathematical property is being applied when *undoing* addition?
  - *Additive inverse*

- What mathematical property allows us to perform an operation (i.e., do the same thing) on both sides of the equation?
  - *Addition and multiplication properties of equality*
- How are the addition and multiplication properties of equality applied?
  - *The problem is an equation which means $A = B$. If a number is added or multiplied to both sides, then the resulting sum or product are equal to each other.*

Exercise (5 minutes)

Exercise

The cost of a babysitting service on a cruise is $\$10$ for the first hour and $\$12$ for each additional hour. If the total cost of babysitting baby Aaron was $\$58$, how many hours was Aaron at the sitter?

Algebraic Solution

h = number of additional hours

$$12h + 10 = 58$$
$$12h + 10 - 10 = 58 - 10$$
$$12h + 0 = 48$$
$$\left(\frac{1}{12}\right)(12h) = (48)\left(\frac{1}{12}\right)$$
$$1h = 4$$
$$h = 4$$

$1 + 4 = 5$

Tape Diagram

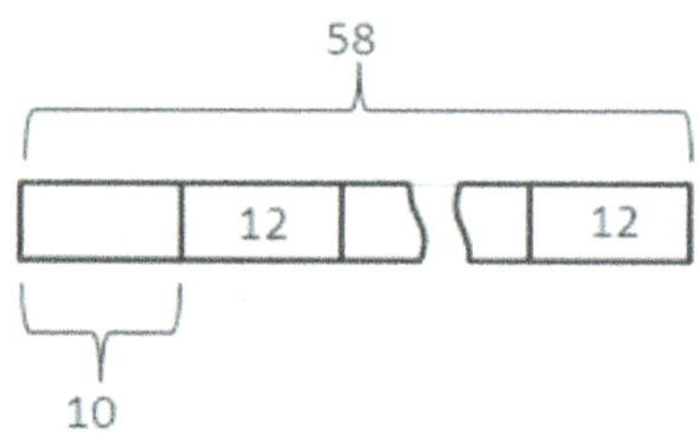

$10 + 12 + 12 = 34$ *(not enough, need 58)*

10	12	12	12

$10 + 12 + 12 + 12 = 46$ *(not enough, need 58)*

10	12	12	12	12

$10 + 12 + 12 + 12 + 12 = 58$

$58 - 10 = 48$

$48 \div 12 = 4$

Aaron was with the babysitter for 5 hours.

MP.4

- How can a tape diagram be used to model this problem?
  - *A tape diagram can be set up to show each hour and the cost associated with that hour. The total is known, so the sum of each column in the tape diagram can be calculated until the total is obtained.*

MP.1

- How is the tape diagram for this problem similar to the tape diagrams used in the previous activity?
  - *In all the problems, the total was given.*

EUREKA MATH™

MP.1

- How is the tape diagram for this problem different than the tape diagrams used in the Exploratory Challenge?
  - *In the Exploratory Challenge, we knew how many units there were, such as days, hours, and people. What was obtained was the amount for one of those units. In this tape diagram, we don't know how many units there are but rather how much each unit represents. Therefore, to solve, we calculate the sum and increase the number of units until we obtain the given sum.*

Closing (3 minutes)

- How does modeling the sequence of operations with a tape diagram help to solve the same problem algebraically?
  - *The tape diagrams provide a visual model to demonstrate what we do when solving the problem algebraically.*
- What are the mathematical properties, and how are they used in finding the solution of a linear equation containing parentheses?
  - *If a linear equation has parentheses, you can either solve using the distributive property or the multiplicative inverse.*

Lesson Summary

Tape diagrams **can be used to model and identify the sequence of operations to find a solution algebraically.**

The goal in solving equations algebraically is to isolate the variable.

The process of doing this requires *undoing* addition or subtraction to obtain a 0 and *undoing* multiplication or division to obtain a 1. The additive inverse and multiplicative inverse properties are applied to get the 0 (the additive identity) and 1 (the multiplicative identity).

The addition and multiplication properties of equality are applied because in an equation, $A = B$, when a number is added or multiplied to both sides, the resulting sum or product remains equal.

Exit Ticket (7 minutes)

Complete one of the problems. Solve by modeling the solution with a tape diagram, and write and solve an algebraic equation.

Name ______________________________ Date ______________

Lesson 17: Comparing Tape Diagram Solutions to Algebraic Solutions

Exit Ticket

1. Eric's father works two part-time jobs, one in the morning and one in the afternoon, and works a total of 40 hours each 5-day workweek. If his schedule is the same each day, and he works 3 hours each morning, how many hours does Eric's father work each afternoon?

2. Henry is using a total of 16 ft. of lumber to make a bookcase. The left and right sides of the bookcase are each 4 ft. high. The top, bottom, and two shelves are all the same length, labeled S. How long is each shelf?

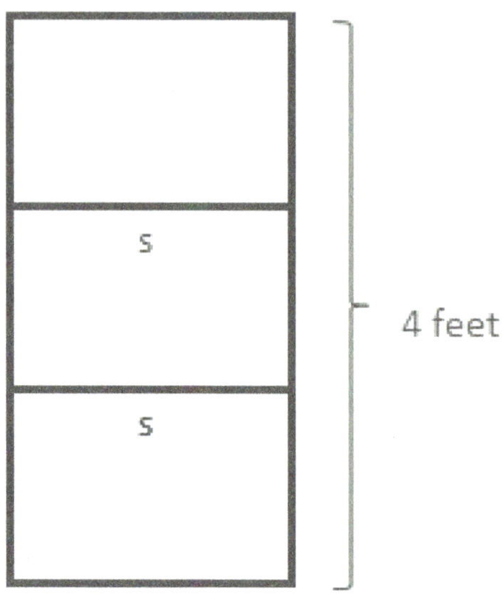

Exit Ticket Sample Solutions

1. Eric's father works two part-time jobs, one in the morning and one in the afternoon, and works a total of 40 hours each 5-day workweek. If his schedule is the same each day, and he works 3 hours each morning, how many hours does Eric's father work each afternoon?

Algebraic Equation & Solution

Number of afternoon hours: a

Number of morning hours: 3

$$5(a+3)=40$$
$$5a+15-15=40-15$$
$$5a+0=25$$
$$\left(\frac{1}{5}\right)5a=25\left(\frac{1}{5}\right)$$
$$a=5$$

OR

$$5(a+3)=40$$
$$\left(\frac{1}{5}\right)5(a+3)=40\left(\frac{1}{5}\right)$$
$$a+3=8$$
$$a+3-3=8-3$$
$$a=5$$

Tape Diagram

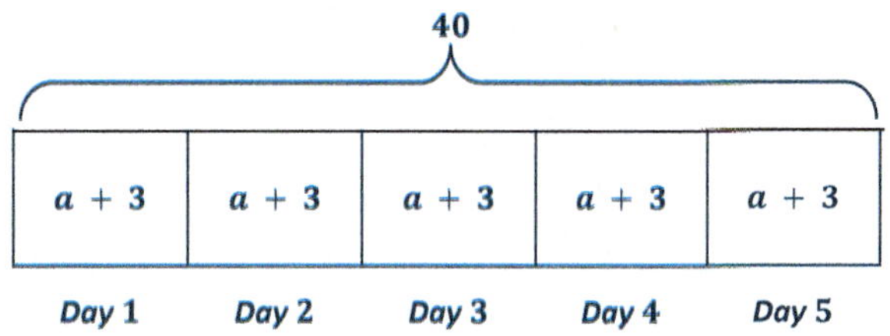

$5(3)=15$ $40-15=25$ $25\div 5=5$

Eric's father works 5 *hours in the afternoon.*

2. Henry is using a total of 16 ft. of lumber to make a bookcase. The left and right sides of the bookcase are each 4 ft. high. The top, bottom, and two shelves are all the same length, labeled S. How long is each shelf?

Algebraic Equation & Solution

Shelves: s ft.

Sides: 8 ft.

$$4s+8=16$$
$$4s+8-8=16-8$$
$$4s+0=8$$
$$\left(\frac{1}{4}\right)4s=8\left(\frac{1}{4}\right)$$
$$1s=2$$
$$s=2$$

Tape Diagram

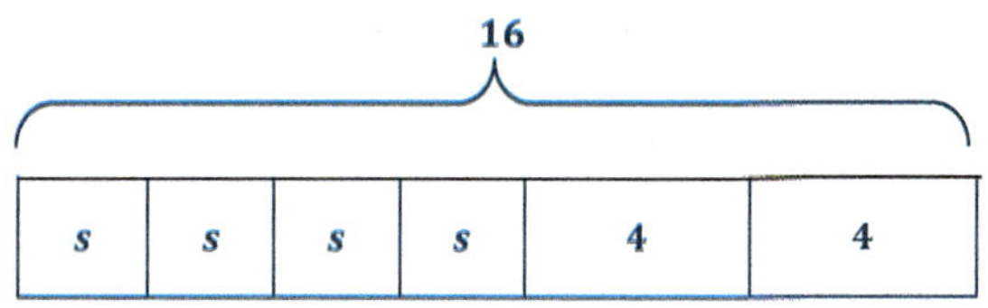

$2(4)=8$ $16-8=8$ $8\div 4=2$

Each shelf is 2 *ft. long.*

Problem Set Sample Solutions

1. A taxi cab in Myrtle Beach charges $\$2$ per mile and $\$1$ for every person. If a taxi cab ride for two people costs $\$12$, how far did the taxi cab travel?

Algebraic Equation & Solution

Number of miles: m

People: 2

$$2m + 2 = 12$$
$$2m + 2 - 2 = 12 - 2$$
$$2m + 0 = 10$$
$$\left(\frac{1}{2}\right)2m = 10\left(\frac{1}{2}\right)$$
$$1m = 5$$
$$m = 5$$

Tape Diagram

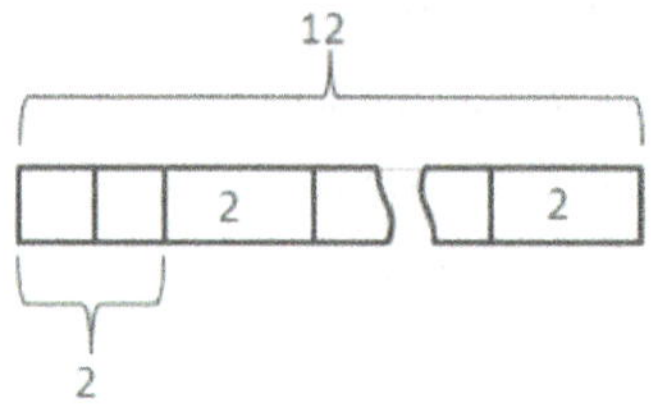

$12 - 2 = 10$ $\quad$ $10 \div 2 = 5$

The taxi cab traveled 5 miles.

2. Heather works as a waitress at her family's restaurant. She works 2 hours every morning during the breakfast shift and returns to work each evening for the dinner shift. In the last four days, she worked 28 hours. If Heather works the same number of hours every evening, how many hours did she work during each dinner shift?

Algebraic Equation & Solution

Number of morning hours: 2

Number of evening hours: e

$$4(e + 2) = 28$$
$$4e + 8 - 8 = 28 - 8$$
$$4e + 0 = 20$$
$$\left(\frac{1}{4}\right)4e = 20\left(\frac{1}{4}\right)$$
$$1e = 5$$
$$e = 5$$

OR

$$\left(\frac{1}{4}\right)4(e + 2) = 28\left(\frac{1}{4}\right)$$
$$e + 2 = 7$$
$$e + 2 - 2 = 7 - 2$$
$$e = 5$$

Tape Diagram

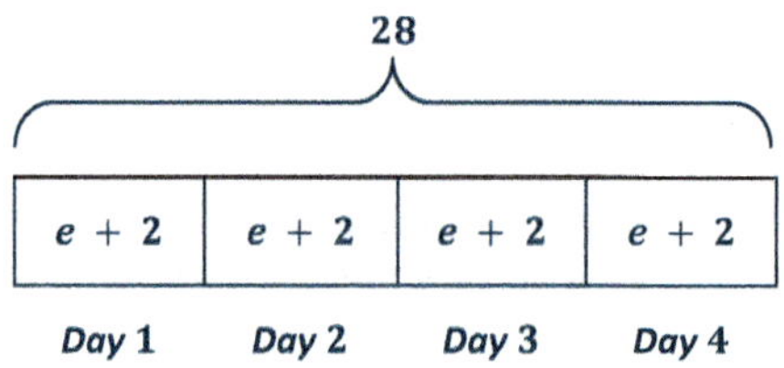

$4e$ $\quad$ $28 - 8 = 20$

$4(2) = 8$ $\quad$ $20 \div 4 = 5$

Heather worked 5 hours in the evening.

3. Jillian exercises 5 times a week. She runs 3 miles each morning and bikes in the evening. If she exercises a total of 30 miles for the week, how many miles does she bike each evening?

Algebraic Equation & Solution

Run: 3 mi.

Bikes: b mi.

$$5(b+3)=30$$
$$5b+15-15=30-15$$
$$5b+0=15$$
$$\left(\frac{1}{5}\right)5b=15\left(\frac{1}{5}\right)$$
$$1b=3$$
$$b=3$$

OR

$$\left(\frac{1}{5}\right)(b+3)=30\left(\frac{1}{5}\right)$$
$$b+3=6$$
$$b+3-3=6-3$$
$$b=3$$

Tape Diagram

30

$b+3$	$b+3$	$b+3$	$b+3$	$b+3$
Day 1	*Day* 2	*Day* 3	*Day* 4	*Day* 5

$5(3)=15$ $\quad 30-15=15$

$15\div 5=3$

Jillian bikes 3 *miles every evening.*

4. Marc eats an egg sandwich for breakfast and a big burger for lunch every day. The egg sandwich has 250 calories. If Marc has $5,250$ calories for breakfast and lunch for the week in total, how many calories are in one big burger?

Algebraic Equation & Solution

Egg Sandwich: 250 cal.

Hamburger: m cal

$$7(m+250)=5,250$$
$$7m+1,750-1750=5250-1750$$
$$7m+0=3,500$$
$$\left(\frac{1}{7}\right)7m=3,500\left(\frac{1}{7}\right)$$
$$1m=500$$
$$m=500$$

OR

$$\left(\frac{1}{7}\right)7(m+250)=\left(\frac{1}{7}\right)5,250$$
$$m+250=750$$
$$m+250-250=750-250$$
$$m=500$$

Tape Diagram

$5,250$

$m+250$	$m+250$	$m+250$	$m+250$	$m+250$	$m+250$	$m+250$
Day 1	*Day* 2	*Day* 3	*Day* 4	*Day* 5	*Day* 6	*Day* 7

$7(250)=1,750$

$5,250-1,750=3,500$

$3,500\div 7=500$

Each hamburger has 500 *calories.*

5. Jackie won tickets playing the bowling game at the local arcade. The first time, she won 60 tickets. The second time, she won a bonus, which was 4 times the number of tickets of the original second prize. Altogether she won 200 tickets. How many tickets was the original second prize?

Algebraic Equation & Solution

First Prize: 60 *tickets*

Second Prize: p *tickets*

$$4p + 60 = 200$$

$$4p + 60 - 60 = 200 - 60$$

$$4p + 0 = 140$$

$$\left(\frac{1}{4}\right)4p = 140\left(\frac{1}{4}\right)$$

$$1p = 35$$

$$p = 35$$

Tape Diagram

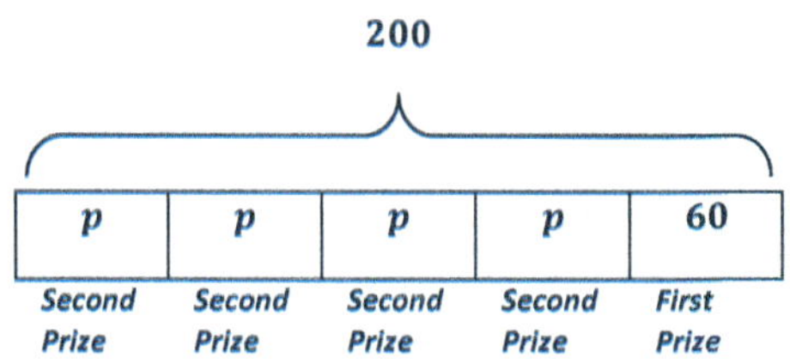

$4p$

$200 - 60 = 140$

$140 \div 4 = 35$

The original second prize was 35 *tickets.*

Exploratory Challenge Scenarios

Scenario 1

During one rainy day on the vacation, the entire family decided to go watch a matinee movie in the morning and a drive-in movie in the evening. The price for a matinee movie in the morning is different than the cost of a drive-in movie in the evening. The tickets for the matinee morning movie cost $\$6$ each. How much did each person spend that day on movie tickets if the ticket cost for each family member was the same? What was the cost for a ticket for the drive-in movie in the evening?

Scenario 2

For dinner one night, the family went to the local pizza parlor. The cost of a soda was $\$3$. If each member of the family had a soda and one slice of pizza, how much did one slice of pizza cost?

Scenario 3

One night, John, Louie, and Bonnie went to see the local baseball team play a game. They each bought a game ticket and a hat that cost $\$10$. How much was each ticket to enter the ballpark?

Scenario 4

While John, Louie, and Bonnie went to see the baseball game, Ag and Missy went shopping. They bought a T-shirt for each member of the family and bought two pairs of sandals that cost $\$10$ a pair. How much was each T-shirt?

Scenario 5

The family flew in an airplane to their vacation destination. Each person had to have his own ticket for the plane and also pay $\$25$ in insurance fees per person. What was the cost of one ticket?

Scenario 6

While on vacation, the family rented a car to get them to all the places they wanted to see for five days. The car costs a certain amount each day, plus a one-time insurance fee of $\$50$. How much was the daily cost of the car (not including the insurance fees)?

Scenario 7

The family decided to stay in a motel for four nights. The motel charges a nightly fee plus $\$60$ in state taxes. What is the nightly charge with no taxes included?

Lesson 18: Writing, Evaluating, and Finding Equivalent Expressions with Rational Numbers

Student Outcomes

- Students create equivalent forms of expressions in order to see structure, reveal characteristics, and make connections to context.
- Students compare equivalent forms of expressions and recognize that there are multiple ways to represent the context of a word problem.
- Students write and evaluate expressions to represent real-world scenarios.

Classwork

Exercise 1 (15 minutes)

MP.1 & MP.2 Students work with a partner or small group to determine the best cell phone plan when given the different prices and options for three different companies. Students are required to write an expression to represent each plan and evaluate each expression to determine which plan is most economical.

Prior to the exercise, recall the description of an expression:

- An expression can be a number or a letter (which can be raised to a whole number exponent) that represents a number. What are some examples?
  - $x, 3, x^2$
- An expression can be the product whose factors are any one of the entities described above. What are some examples?
  - $3 \cdot 2$
- An expression can also be the sum and/or difference of the products described above. What are some examples?
  - $3 \cdot 2 + x - 2$

Exercise 1

John's father asked him to compare several different cell phone plans and identify which plan will be the least expensive for the family. Each phone company charges a monthly fee, but this fee does not cover any services: phone lines, texting, or internet access. Use the information contained in the table below to answer the following questions.

Cell Phone Plans

Name of Plan	Monthly Fee (Includes $1,500$ shared minutes)	Price per Phone Line x	Price per Line for Unlimited Texting y	Price per Line for Internet Access z
Company A	$\$70$	$\$20$	$\$15$	$\$15$
Company B	$\$90$	$\$15$	$\$10$	$\$20$
Company C	$\$200$	$\$10$	included in monthly fee	included in monthly fee

All members of the family may not want identical plans; therefore, we will let x represent the number of phone lines, y represent the number of phone lines with unlimited texting, and z represent the number of phone lines with internet access.

Expression

Company A $70 + 20x + 15y + 15z$

Company B $90 + 15x + 10y + 20z$

Company C $200 + 10x$

Using the expressions above, find the cost to the family of each company's phone plan if:

a. Four people want a phone line, four people want unlimited texting, and the family needs two internet lines.

Company A	Company B	Company C
$70 + 20x + 15y + 15z$	$90 + 15x + 10y + 20z$	$200 + 10x$
$70 + 20(4) + 15(4) + 15(2)$	$90 + 15(4) + 10(4) + 20(2)$	$200 + 10(4)$
$70 + 80 + 60 + 30$	$90 + 60 + 40 + 40$	$200 + 40$
240	230	240

Which cell phone company should John's family use? Why?

The family should choose Company B since it is cheaper than the others for the given values.

b. Four people want a phone line, four people want unlimited texting, and all four people want internet lines.

Company A	Company B	Company C
$70 + 20x + 15y + 15z$	$90 + 15x + 10y + 20z$	$200 + 10x$
$70 + 20(4) + 15(4) + 15(4)$	$90 + 15(4) + 10(4) + 20(4)$	$200 + 10(4)$
$70 + 80 + 60 + 60$	$90 + 60 + 40 + 80$	$200 + 40$
270	270	240

Which cell phone company should John's family use? Why?

The family should choose Company C since it is cheaper than the other companies for the given values.

c. Two people want a phone line, two people want unlimited texting, and the family needs two internet lines.

Company A	Company B	Company C
$70 + 20x + 15y + 15z$	$90 + 15x + 10y + 20z$	$200 + 10x$
$70 + 20(2) + 15(2) + 15(2)$	$90 + 15(2) + 10(2) + 20(2)$	$200 + 10(2)$
$70 + 40 + 30 + 30$	$90 + 30 + 20 + 40$	$200 + 20$
170	180	220

Which cell phone company should John's family use? Why?

The family should choose Company A since it is cheaper than the other companies for the given values.

Questions to follow exercise:

- Why is there no equal sign in the expressions?
  - *There is no equal sign because we are writing an expression and not an equation.*
- Each plan charges for four different options, yet there are only three variables in each expression. Why is this?
  - *There are only three variables used because the monthly fee is added to the options regardless of how many lines are purchased.*
- In order to have a plan, a customer would have at least one phone line. What would be the minimum cost for each plan?
  - *The minimum cost for each plan would be the monthly fee and one phone line. Company A would be* $\$90$*, Company B would be* $\$105$*, and Company C would be* $\$210$*.*
- What role did the expression play in your decision-making process?
  - *Writing an expression allowed us to evaluate and compare the different companies for many different values of each variable.*
- Describe the process you used to arrive at the total cost of each plan?
  - *First, the given values for each variable are substituted into the expression, so every variable is replaced with the corresponding numerical value. After that, you do the arithmetic following order of operations.*

Exercise 2 (10 minutes)

Students continue to write and evaluate expressions from real-world problems, but also identify equivalent expressions during the process.

MP.1 & MP.2

In the same groups as Exercise 1, have students first individually read through the following problem, write an expression, and evaluate the expression. Once completed, have students compare their results with those of their group. Once all members of the group agree upon the correct answer, students should compare their solutions, looking for similarities and differences among the various methods used. If there are any differences, have them discuss what they were. As a large group, have a discussion about the different possible ways of obtaining the same solution. If any person or group obtained the solution in different ways, have them explain their process to the class.

Exercise 2

Three friends went to the movies. Each purchased a medium-sized popcorn for p dollars and a small soft drink for s dollars.

a. **Write the expression that represents the total amount of money (in dollars) the three friends spent at the concession stand.**

   $3(p+s)$

b. **If the concession stand charges $\$6.50$ for a medium-sized popcorn and $\$4.00$ for a small soft drink, how much did the three friends spend on their refreshments altogether?**

   One possible solution is shown here; more solution methods are shown in the discussion that follows.

   $3(p+s)$

   $3(6.50+4)$

   $3(10.50)$

   31.50

   They spent $\$31.50$*.*

Questions to follow the exercise:

- What information did you use to write the expression?
  - *You needed to know what the variables were and what they represented. You also needed to know or figure out how many people were getting each item.*
- John created the expression $3p + 3s$ to represent the total cost of the refreshments while Sally used the expression $3(p + s)$. Are they both correct? If so, what do the expressions tell us about the two different ways in which John and Sally calculate the cost of the refreshments?
  - *Yes, both expressions are correct. John calculated the cost of three drinks and the cost of three popcorns and added them together. Sally calculated the amount each friend spent and then multiplied by the number of friends.*
- Compare the four samples of student work. What are the differences among the four methods? (This is where groups or individuals can share their work if it matches any of these.)

Student 1:	Student 2:	Student 3:	Student 4:
$3(p + s)$	$3(p + s)$	$3(p + s)$	$3(p + s)$
$3(6.50 + 4)$	$3(6.50 + 4)$	$3(6.50 + 4)$	$(p + s) + (p + s) + (p + s)$
$3(10.50)$	$3(10.50)$	$3(6.50) + 3(4)$	$(6.50 + 4) + (6.50 + 4) + (6.50 + 4)$
31.50	$3(10 + 0.50)$	$19.50 + 12$	$10.50 + 10.50 + 10.50$
	$30 + 1.50$	31.50	31.50
	31.50		

While discussing the differences in the four methods, clearly describe that the methods above are beginning with the same expression but that each method demonstrates a different way of evaluating the same expression.

- The next time the three friends went to the movies, they each purchased a small-sized soft drink but decided to share one medium-sized popcorn. Write the expression that describes the amount the group spent at the concession stand. How does this expression differ from the one you created before?
  - *Expression: $3s + p$. Only the value of the soda would be multiplied by 3 and not that of the popcorn. There would be no need to apply the distributive property.*

Exercise 3 (10 minutes)

Exercise 3

Complete the table below by writing equivalent expressions to the given expression and evaluating each expression with the given values.

Equivalent Expressions			
EXAMPLE: Evaluate $x = 2$, $y = -1$	$4(x + 2y)$ $4(2 + 2(-1))$ $4(0)$ 0	$4x + 8y$ $4(2) + 8(-1)$ $8 + (-8)$ 0	$4x + 4y + 4y$ $4(2) + 4(-1) + 4(-1)$ $8 + (-4) + (-4)$ 0
1. Evaluate $y = 1$	$5(3 - 4y)$ $5(3 - 4(1))$ $5(3 - 4)$ $5(-1)$ -5	$15 - 20y$ $15 - 20(1)$ $15 - 20$ $15 + (-20)$ -5	$15 - 10y - 10y$ $15 - 10(1) - 10(1)$ $15 - 10 - 10$ $5 - 10$ $5 + (-10)$ -5
2. Evaluate $x = 5$, $y = -2$	$-3x + 12y$ $-3(5) + 12(-2)$ $-15 + (-24)$ -39	$3(-1x + 4y)$ $3(-1(5) + 4(-2))$ $3(-5 + (-8))$ $3(-13)$ -39	$-x - x - x + 4y + 4y + 4y$ $-5 - 5 - 5 + 4(-2) + 4(-2) + 4(-2)$ $-5 + (-5) + (-5) + (-8) + (-8) + (-8)$ -39
3. Evaluate $x = -\frac{1}{2}$, $y = 1$	$8x - 6y$ $8\left(-\frac{1}{2}\right) - 6(1)$ $-4 - 6$ $-4 + (-6)$ -10	$2(4x - 3y)$ $2\left(4\left(-\frac{1}{2}\right) - 3(1)\right)$ $2(-2 - 3)$ $2(-2 + (-3))$ $2(-5)$ -10	$-2x + 10x - 6y$ $-2\left(-\frac{1}{2}\right) + 10\left(-\frac{1}{2}\right) - 6(1)$ $1 - 5 - 6$ $1 + (-5) + (-6)$ -10

Questions and discussion to follow the exercise:

- Looking at the equivalent expressions you created, can you see any benefit to using one over the other?
  - *Answers will vary.*
- Would it matter which equivalent expression you use if you were asked to evaluate the expression?
  - *No, the expressions are equivalent because when evaluated, all of the expressions equal the same rational number.*
- For each equivalent expression written in the table, discuss with your elbow partner why each expression is equivalent.

Provide time for students to discuss; then have them share their responses with the class.

- *Sample explanations should include students applying the distributive property.*

Have students identify a context that the expression could be modeling with their partners.

Closing (5 minutes)

- What is an expression? Describe the steps to evaluating it.
  - *An expression is a number or letter (that may have a whole number exponent) that represents a number. Expressions can also be products or sums.*
- To evaluate an expression, substitute numerical values into the expression and then follow order of operations to solve.
- How do you determine if expressions are equivalent?
  - *Equivalent expressions have the same numerical value for every number substituted into the problem.*

Lesson Summary

- **An expression is a number or a letter, which can be raised to a whole number exponent. An expression can be a product whose factors are any one of the entities described above. An expression can also be the sum or difference of the products described above.**
- **To evaluate an expression, replace each variable with its corresponding numerical value. Using order of operations, the expression can be written as a single numerical value.**
- **When numbers are substituted into all the letters in an expression and the results are the same, then the expressions are equivalent.**

Exit Ticket (5 minutes)

Name ______________________________ Date ______________

Lesson 18: Writing, Evaluating, and Finding Equivalent Expressions with Rational Numbers

Exit Ticket

Bradley and Louie are roommates at college. At the beginning of the semester, they each paid a security deposit of A dollars. When they move out, their landlord will deduct from this deposit any expenses (B) for excessive wear and tear and refund the remaining amount. Bradley and Louie will share the expenses equally.

- Write an expression that describes the amount each roommate will receive from the landlord when the lease expires.
- Evaluate the expression using the following information: Each roommate paid a $125 deposit, and the landlord deducted $50 total for damages.

Exit Ticket Sample Solutions

Bradley and Louie are roommates at college. At the beginning of the semester, they each paid a security deposit of A dollars. When they move out, their landlord will deduct from this deposit any expenses (B) for excessive wear and tear and refund the remaining amount. Bradley and Louie will share the expenses equally.

- Write an expression that describes the amount each roommate will receive from the landlord when the lease expires.
- Evaluate the expression using the following information: Each roommate paid a $\$125$ deposit, and the landlord deducted $\$50$ total for damages.

Deposit each person paid: A

Total damages: B

Each roommate receives: $A - \frac{B}{2}$

$A = 125, B = 50$

$A - \frac{B}{2}$

$125 - \frac{50}{2}$

$125 - 25$

100

Problem Set Sample Solutions

1. Sally is paid a fixed amount of money to walk her neighbor's dog every day after school. When she is paid each month, she puts aside $\$20$ to spend and saves the remaining amount. Write an expression that represents the amount Sally will save in 6 months if she earns m dollars each month. If Sally is paid $\$65$ each month, how much will she save in 6 months?

$m =$ *monthly pay*

$6(m - 20)$

$6m - 120$

For $m = 65$

$6(m - 20)$	*or*	$6(m - 20)$
$6(65 - 20)$		$6(65 - 20)$
$6(45)$		$390 - 120$
270		270

Sally will save $\$270$ *in* 6 *months.*

2. A football team scored 3 touchdowns, 3 extra points, and 4 field goals.

   a. Write an expression to represent the total points the football team scored.

      t = number of points for a touchdown.

      e = number of points for the extra point.

      f = number of points for a field goal.

      $3t + 3e + 4f$

   b. Write another expression that is equivalent to the one written above.

      Answers may vary. Sample response: $3t + 3e + 2f + 2f$

   c. If each touchdown is worth 6 points, each extra point is 1 point, and each field goal is 3 points, how many total points did the team score?

      $3t + 3e + 4f$

      $3(6) + 3(1) + 4(3)$

      $18 + 3 + 12$

      33

3. Write three other expressions that are equivalent to $8x - 12$.

   Answers may vary.

   $4(2x - 3)$

   $6x + 2x - 12$

   $8(x - 1) - 4$

   $-12 + 8x$

4. Profit is defined as earnings less expenses (earnings − expenses). At the local hot-air balloon festival, the Ma & Pops Ice Cream Truck sells ice cream pops, which cost them $\$0.75$ each but are sold for $\$2$ each. They also paid $\$50$ to the festival's organizers for a vendor permit. The table below shows the earnings, expenses, and profit earned when 50, 75, and 100 ice cream pops were sold at the festival.

Number of Pops Sold	Earnings	Expenses	Profit
50	$50(2) = 100$	$50(0.75) + 50$ $37.5 + 50 = 87.5$	$100 - 87.5 = 12.50$
75	$75(2) = 150$	$75(0.75) + 50$ $56.25 + 50 = 106.25$	$150 - 106.25 = 43.75$
100	$100(2) = 200$	$100(0.75) + 50$ $75 + 50 = 125$	$200 - 125 = 75$

   a. Write an expression that represents the profit (in dollars) Ma & Pops earned by selling ice cream pops at the festival.

      p represents the number of pops sold.

      $2p - 0.75p - 50$

b. **Write an equivalent expression.**

$1.25p - 50$

c. **How much of a profit did Ma & Pops Ice Cream Truck make if it sold 20 ice cream pops? What does this mean? Explain why this might be the case.**

$1.25p - 50$

$1.25(20) - 50$

$25 - 50$

-25

They did not make any money; they lost $\$25$. A possible reason is it could have been cold or rainy and people were not buying ice cream.

d. **How much of a profit did Ma & Pops Ice Cream Truck make if it sold 75 ice cream pops? What does this mean? Explain why this might be the case.**

$1.25p - 50$

$1.25(75) - 50$

$93.75 - 50$

43.75

They made a profit of $\$43.75$. Possible reasons are the weather could have been warmer and people bought the ice cream, or people just like to eat ice cream no matter what the weather is.

Lesson 19: Writing, Evaluating, and Finding Equivalent Expressions with Rational Numbers

Student Outcomes

- Students create equivalent forms of expressions in order to see structure, reveal characteristics, and make connections to context.
- Students compare equivalent forms of expressions and recognize that there are multiple ways to represent the context of a word problem.
- Students write and evaluate expressions to represent real-world scenarios.

Lesson Notes

Students should round to the nearest penny whenever necessary. Make sure this is emphasized throughout the lesson.

Classwork

Example 1 (10 minutes): Tic-Tac-Toe Review

Begin by having students play an equivalent expression Tic-Tac-Toe game. Have students randomly fill in the 9 spots on their game boards with an expression from the student list of 10. Once students have their game boards filled in, show them an expression from the teacher list. Have students then find and mark (with an X) all equivalent expressions on their game boards. A student wins the game by getting 3 in a row.

Suggestion: Go through all of the expressions for practice even if the game is won before the end. The expression $1(x+2) + 2(x-2)$ from the teacher's list is equivalent to $3x-2$, which is not on the students' game board. Discuss with students why and how $3x-2$ is not the same as $3(x-2)$.

Example 1: Tic-Tac-Toe Review

Fill in the 9 spaces with one expression from the list below. Use one expression per space. You will use 9 of the expressions:

$12 - 4x$

$8x + 4 - 12x$

$8\left(\frac{1}{2}x - 2\right)$

$12 - 6x + 2x$

$-4x + 4$

$x - 2 + 2x - 4$

$4x - 12$

$4(x - 4)$

$3(x - 2)$

$0.1(40x) - \frac{1}{2}(24)$

G7-M2-B2-v1.3.0_05/04/15

Teacher List

$2x + 2(x - 6)$

$4x - 16$

$1(x + 2) + 2(x - 2)$

$4(3 - x)$

$4(2x + 1) - 12x$

$3x - 6$

Example 2 (12 minutes)

Students complete the first row by using their knowledge of percents and discounts to find the discount amount and new price when the original price is given. Students then write a numerical and/or equivalent expression to find the new price of different items whose original price is given. The teacher leads the discussion in showing students how the problem can be solved both by arithmetic, as well as visually, using a tape diagram. Students extend this by creating expressions that combine discounts (and include sales tax using whichever approach they prefer).

Example 2

Original Price (100%)	Discount Amount $(20\%$ Off)	New Price (Pay 80%)	Expression
$\mathbf{100}$	$\mathbf{100(0.20) = 20}$	$\mathbf{100 - 20 = 80}$	$\mathbf{100 - 100(0.20)}$ $\mathbf{100(1 - 0.20)}$ $\mathbf{100(0.80)}$
$\mathbf{50}$	$\mathbf{50(0.20) = 10}$	$\mathbf{50 - 10 = 40}$	$\mathbf{50 - 50(0.20)}$ $\mathbf{50(1 - 0.20)}$ $\mathbf{50(0.80)}$
$\mathbf{28}$	$\mathbf{28(0.20) = 5.60}$	$\mathbf{28 - 5.60 = 22.40}$	$\mathbf{28 - 28(0.20)}$ $\mathbf{28(1 - 0.20)}$ $\mathbf{28(0.80)}$
$\mathbf{14.50}$	$\mathbf{14.50(0.20) = 2.90}$	$\mathbf{14.50 - 2.90 = 11.60}$	$\mathbf{14.50 - 14.50(0.20)}$ $\mathbf{14.50(1 - 0.20)}$ $\mathbf{14.50(0.80)}$
$\boldsymbol{x}$	$\boldsymbol{x(0.20) = 0.20x}$	$\boldsymbol{x - 0.20x}$	$\boldsymbol{x - 0.20x}$ $\boldsymbol{x(1 - 0.20)}$ $\boldsymbol{x(0.80)}$

Discussion

A discount is an amount that is subtracted from the original price.

- If you know the original cost of an item, how do you find the discount amount by using a picture and by using arithmetic?
  - *Answers will vary.*

The intent is for students to complete the first row before the teacher leads a discussion on how to find the discount with both a picture and arithmetic. After that, students may use whichever method they prefer. Some students may choose to calculate 10% of the total and then double it to find 20%.

- Picture: 20% off $\$100$.

$\frac{20}{100} = \frac{1}{5}$ *Make a tape diagram and break the whole into 5 parts, each part representing 20%.*

Then divide the total amount of money into 5 parts. The discount is the amount represented in one of the parts; the amount paid is the remaining parts.

$\$100$

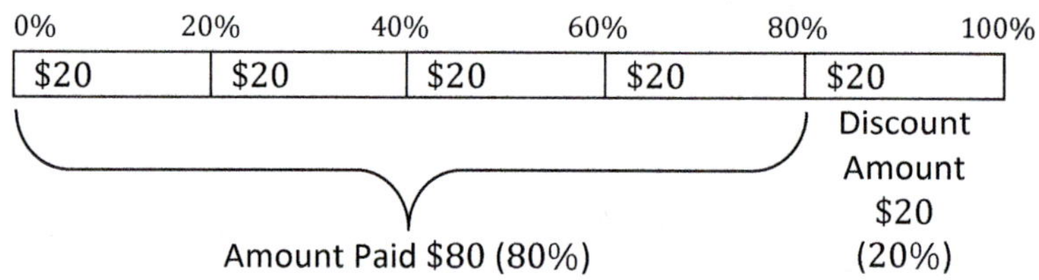

$\$50$

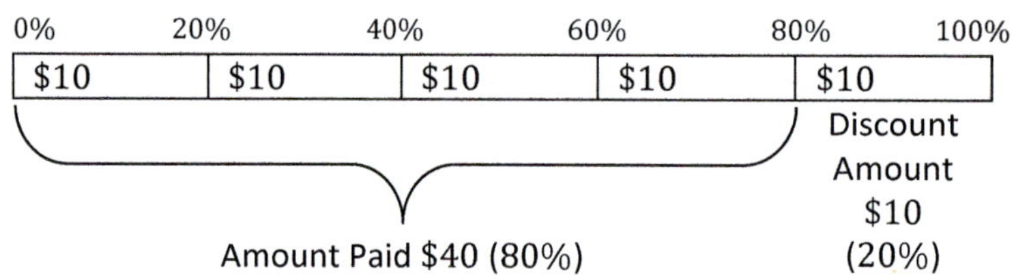

$\$28$

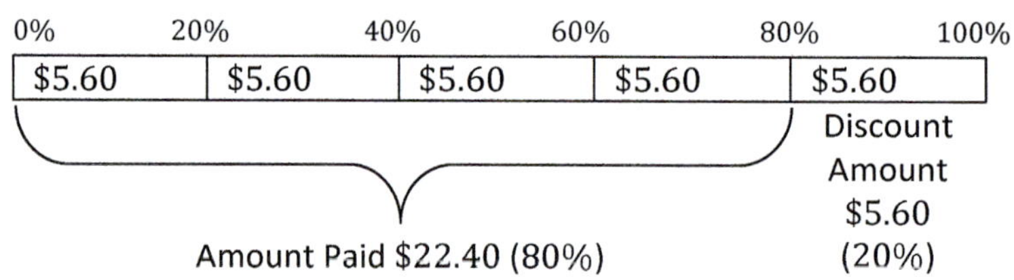

$\$14.50$

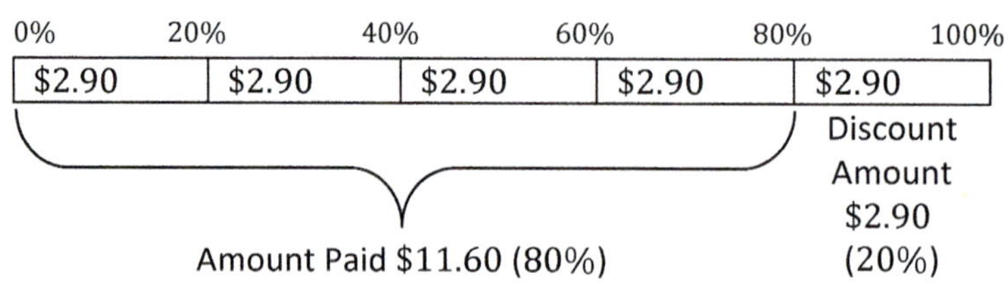

Arithmetic: Calculate the amount of discount that corresponds to the discount percent using any method. Then subtract this value from the original amount.

- How do you set up a tape diagram for a percent that is not a factor of 100, for example 30%?
  - *Determine the greatest common factor of the percent and 100. Divide 100 by the greatest common factor, and that will determine into how many parts to break the tape diagram.*
  - *Since 30 is not a factor of 100, find the greatest common factor of 30 and 100. The greatest common factor of 30 and 100 is 10. Therefore, when 100 is divided by the greatest common factor of 10, the result of 10 indicates how many parts into which to break the tape diagram.*
- What is the process to find a percent of a number without using a tape diagram?
  - *Multiply the whole by the percent as a fraction out of 100, or multiply the whole by the percent, written as a decimal.*
- Under what circumstances would you prefer to use a tape diagram to help you calculate the percent of a number?
  - *Finding the percent of a number using arithmetic is sometimes quicker than using a tape diagram. Using a tape diagram would be most beneficial when the percent and 100 have a greatest common factor and when the GCF isn't so small that it divides the tape diagram into numerous parts.*
- When the original price is not known, how can an expression be used to represent the new price?
  - *When the original price is unknown, it can be represented by a variable such as x. To write an expression that represents the new price, the discount amount must be subtracted from the original amount. The expression can then be written as an equivalent expression.*
- When a discount of 20% is being deducted, what percent is being paid? How do you know?
  - *The amount being paid would be 80%. We know this because an item not on sale represents 100%. If there is a discount of 20%, then the overall price would be 20% less than the original 100%. To find this, subtract 20% from 100%, and the difference is the percent that is paid.*
- How is $x - 0.2x = 0.8x$?
  - *When the expression $x - 0.2x$ is written as an equivalent expression, you know that x represents $1x$, and when you subtract $0.2x$ from $1x$ the result is $0.8x$.*
- Describe the meaning of $x - 0.2x = 0.8x$ in the context of the problem.
  - *The original price of the item is unknown, represented by x. If the item is on sale for 20%, then the percent that is paid is 80%. $x - 0.20x$ represents the original price less the discount amount, which will equal the new price. The new price is the price that is paid, which is 80% of the original cost, which is represented as $0.8x$.*

Example 3 (5 minutes)

Example 3

An item that has an original price of x dollars is discounted 33%.

a. **Write an expression that represents the amount of the discount.**

$0.33x$

b. **Write two equivalent expressions that represent the new, discounted price.**

$x - 0.33x$

$x(1 - 0.33)$

$x(0.67)$

c. **Use one of your expressions to calculate the new, discounted price if the original price was $\$56$.**

$0.67x$

$0.67(56)$

37.52

The new discounted price is $\$37.52$.

d. **How would the expressions you created in parts (a) and (b) have to change if the item's price had increased by 33% instead of decreased by 33%?**

Instead of subtracting $0.33x$*, you would have to add for the increase. The expression would be*

$x + 0.33x$

$1.33x$.

Example 4 (10 minutes)

Discussion

Generate a classroom discussion about a new concept—the concept of sales tax. Discuss what it is, the purpose of it, and how it is calculated.

MP.1 & MP.2 Once students have a general understanding that sales tax is a number added to the cost of an item, and it is found by finding the sales rate (%) of the item and added to the cost, lead students through the second chart, which is an extension of the first.

Example 4

Original Price (100%)	Discount (20% off)	Amount Pay (pay 80%)	Expression	New Price	Sales Tax (8%)	Overall Cost	Expression
100	20	80	$100 - 100(0.20)$ $= 100(0.80)$				
50	10	40	$50 - 50(0.20)$ $= 50(0.80)$				
28	5.60	22.40	$28 - 28(0.20)$ $= 28(0.80)$				
14.50	2.90	11.60	$14.50 - 14.50(0.20)$ $= 14.50(0.80)$				
x	$0.20x$	$x - 0.20x$	$x - 0.20x$ $= 0.80x$				

EUREKA MATH™

- If a tape diagram were used to model the sales tax, into how many parts would the tape diagram need to be broken? Explain how you knew that.
  - *Since the GCF of* 8 *and* 100 *is* 4, *the tape diagram would need to broken into* $\frac{100}{4} = 25$ *parts. This is not the easiest or most efficient way of finding the sales tax.*
- What is 1% of 80?
  - 0.80
- If you can find 1% of 80 easily, how can you use that answer to find 8% of 80?
  - *Multiply by* 8 *because* 1% *multiplied by* 8 *will give* 8%.

    $0.80(8) = 6.40$
  - *Arithmetic:*

    $80 \cdot 0.08 = 6.40$

    $80 + 6.40 = 86.40$

Overall expression:

- What was the expression for the discount?
  - $100(0.80)$
- Using the previous expression, write an expression to determine the amount of the sales tax?
  - $\big(100(0.80)\big)(0.08)$
- Would it change the final price of the item if the sales clerk charged the sales tax first and then discounted the item? Why do you think this is the case?
  - *No, the order wouldn't matter. If the sales tax was calculated first, then the discount would be calculated on both the original price of the item and on the sales tax as well.*
- Describe the process for calculating the final cost of an item, which has been discounted 20% and was sold in a state that has a sales tax of 8%.
  - *Step 1: First take the original amount, and multiply by* 0.20 *to figure out the discount amount.*
  - *Step 2: Use that amount from Step 1, and subtract from the original amount.*
  - *Step 3: Use the new amount from Step 2 (original* $-$ 0.20*(original)), and multiply by* 0.08 *to figure out the sales tax amount.*
  - *Step 4: Use that new amount from Step 3, and add to the discounted price from Step 2.*

- Using the steps you described, write an expression to represent the price paid after a 20% discount and 8% sales tax if the original price was $\$100$. Describe in words what is being found at each step.
  - $100(0.20)$ — *Finding the discount amount*
  - $100 - 100(0.20)$ — *Finding the discount price after 20% is deducted*
  - $(100 - 100(0.20))(0.08)$ — *Finding the sales tax of 8% on the new discounted price*
  - $(100 - 100(0.20)) + 0.08(100 - 100(0.20))$ — *Finding the total paid after finding the discounted price and sales tax on that discounted price and adding them together*
  - $1.08(100(0.80))$
- Using the same steps, write an expression to represent the price paid if the original price is $\$50$ with a 20% discount and 8% sales tax. Describe in words what is being found at each step.
  - $50(0.20)$ — *Finding the discount amount*
  - $50 - 50(0.20)$ — *Finding the discount price after 20% is deducted*
  - $(0.08)(50 - 50(0.20))$ — *Finding the sales tax of 8% on the new discounted price*
  - $(50 - 50(0.20)) + (0.08)(50 - 50(0.20))$ — *Finding the total paid after finding the discounted price and sales tax on that discounted price and adding them together*
  - $1.08(50(0.80))$
- Using the same steps, write an expression to represent the price paid if the original price is $\$28$ with a 20% discount and 8% sales tax. Describe in words what is being found at each step.
  - $28(0.20)$ — *Finding the discount amount*
  - $28 - 28(0.20)$ — *Finding the discount price after 20% is deducted*
  - $(0.08)(28 - 28(0.20))$ — *Finding the sales tax of 8% on the new discounted price*
  - $(28 - 28(0.20)) + 0.08(28 - 28(0.20))$ — *Finding the total paid after finding the discounted price and sales tax on that discounted price and adding them together*
  - $1.08(28(0.80))$
- Using the same steps, write an expression to represent the price paid if the original price is $\$14.50$ with a 20% discount and 8% sales tax. Describe in words what is being found at each step.
  - $14.50(0.20)$ — *Finding the discount amount*
  - $14.50 - 14.50(0.20)$ — *Finding the discount price after 20% is deducted*
  - $(0.08)(14.50 - 14.50(0.20))$ — *Finding the sales tax of 8% on the new discounted price*
  - $(14.50 - 14.50(0.20)) + 0.08(14.50 - 14.50(0.20))$ — *Finding the total paid after finding the discounted price and sales tax on that discounted price and adding them together*
  - $1.08(28(0.80))$

- Using the same steps, write an expression to represent the price paid if the original price is x with a 20% discount and 8% sales tax. Describe in words what is being found at each step.
  - $x(0.20)$ — *Finding the discount amount*
  - $x - x(0.20)$ — *Finding the discount price after 20% is deducted*
  - $(0.08)(x - x(0.20))$ — *Finding the sales tax of 8% on the new discounted price*
  - $(x - x(0.20)) + 0.08(x - x(0.20))$ — *Finding the total paid after finding the discounted price and sales tax on that discounted price and adding them together*
  - $1.08(x(0.80))$

Original Price (100%)	Discount (20% off)	Amount Pay (pay 80%)	Expression	New Price	Sales Tax (8%)	Overall Cost	Expression
100	20	80	$100 - 100(0.20)$ $= 100(0.80)$	80	$80(0.08)$ $= 6.40$	$80 + 6.40$ $= 86.40$	$(100 - 100(0.20)) + 0.08(100 - 100(0.20))$ *or* $1.08(100 - 100(0.20))$ *or* $1.08(100(0.80))$
50	10	40	$50 - 50(0.20)$ $= 50(0.80)$	40	$40(0.08)$ $= 3.20$	$40 + 3.20$ $= 43.20$	$(50 - 50(0.20)) + 0.08(50 - 50(0.20))$ *or* $1.08(50 - 50(0.20))$ *or* $1.08(50(0.80))$
28	5.60	22.40	$28 - 28(0.20)$ $= 28(0.80)$	22.40	$22.40(0.08)$ $= 1.79$	$22.40 + 1.79$ $= 24.19$	$(28 - 28(0.20)) + 0.08(28 - 28(0.20))$ *or* $1.08(28 - 28(0.20))$ *or* $1.08(28(0.80))$
14.50	2.90	11.60	$14.50 - 14.50(0.20)$ $= 14.50(0.80)$	11.60	$11.60(0.08)$ $= 0.93$	$11.60 + 0.93$ $= 12.53$	$(14.50 - 14.50(0.20)) + 0.08(14.50 - 14.50(0.20))$ *or* $1.08(14.50 - 14.50(0.20))$ *or* $1.08(14.50(0.80))$
x	$0.20x$	$x - 0.20x$	$x - 0.20x$ $= 0.80x$	$0.80x$	$0.80x(0.08)$	$0.80x + 0.064x$	$(x - 0.20x) + (x - 0.20x)(0.08)$ *or* $1.08(x - 0.20x)$ *or* $0.80x + 0.064x$ *or* $0.864x$

Discussion

- Describe the meaning of the expression $(x - 0.20x)$?
  - *A number reduced by* 20%
- Describe why $\left((x - 0.20x) + 0.08(x - 0.20)\right)$ is equivalent to $1.08(x - 0.20x)$.
  - *In the first expression,* $(x - 0.20x)$ *gives us the discounted price of the item, and we are adding that value to* 8% *of the discounted price.*
- Describe why $(x - 0.20x) + 0.08(x - 0.20x)$ and $1.08(x - 0.20x)$ are equivalent to $1.08(0.80x)$.
  - *The expression gives* 108% *of the discounted price, which is equivalent to the discounted price of the item plus* 8% *of the discounted price of the item.*

Closing (3 minutes)

- Describe how to write an expression that incorporates the use of multiple percents.
- Describe how expressions with percents can be written as equivalent expressions.

> Lesson Summary
>
> - **Two expressions are equivalent if they yield the same number for every substitution of numbers for the letters in each expression.**
> - **The expression that allows us to find the cost of an item after the discount has been taken and the sales tax has been added is written by representing the discount price added to the discount price multiplied by the sales tax rate.**

Exit Ticket (5 minutes)

Name ______________________________ Date ______________

Lesson 19: Writing, Evaluating, and Finding Equivalent Expressions with Rational Numbers

Exit Ticket

1. Write three equivalent expressions that can be used to find the final price of an item costing g dollars that is on sale for 15% off and charged 7% sales tax.

2. Using all of the expressions, determine the final price for an item that costs $\$75$. If necessary, round to the nearest penny.

3. If each expression yields the same final sale price, is there anything to be gained by using one over the other?

4. Describe the benefits, special characteristics, and properties of each expression.

Exit Ticket Sample Solutions

1. **Write three equivalent expressions that can be used to find the final price of an item costing g dollars that is on sale for 15% off and charged 7% sales tax.**

   $(x - 0.15x) + 0.07(x - 0.15x)$ $\quad$ $1.07(x - 0.15x)$ $\quad$ $1.07(0.85x)$ *or* $0.85(1.07)x$

2. **Using all of the expressions, determine the final price for an item that costs $\$75$. If necessary, round to the nearest penny.**

$x = 75$	$(x - 0.15x) + 0.07(x - 0.15x)$	$1.07(x - 0.15x)$	$1.07(0.85x)$ *or* $0.85(1.07)x$
	$(75 - 0.15(75)) + 0.07(75 - 0.15(75))$	$1.07(75 - 0.15(75))$	$1.07(0.85(75))$
	$63.75 + 0.07(63.75)$	$1.07(63.75)$	$1.07(63.75)$
	$63.75 + 4.46$	68.21	68.21
	68.21		

   The final price of an item that costs $\$75$ is $\$68.21$.

3. **If each expression yields the same final sale price, is there anything to be gained by using one over the other?**

   Using the final two expressions makes the problem shorter and offers fewer areas to make errors. However, all three expressions are correct.

4. **Describe the benefits, special characteristics, and properties of each expression.**

   The second and third expressions collect like terms. The third expression can be written either way using the commutative property of multiplication. The first and second expressions find the discount price first, whereas the third expression is written in terms of percent paid.

Problem Set Sample Solutions

Solve the following problems. If necessary, round to the nearest penny.

1. **A family of 12 went to the local Italian restaurant for dinner. Every family member ordered a drink and meal, 3 ordered an appetizer, and 6 people ordered cake for dessert.**

   a. **Write an expression that can be used to figure out the cost of the bill. Include the definitions for the variables the server used.**

   $d =$ *drink*

   $m =$ *meal*

   $a =$ *appetizer*

   $c =$ *cake*

   $12d + 12m + 3a + 6c$

   b. **The waitress wrote on her ordering pad the following expression: $3(4d + 4m + a + 2c)$.**
   Was she correct? Explain why or why not.

   Yes, she was correct because her expression is equivalent to the expression from part (a). If the distributive property is applied, the expressions would be exact.

c. **What is the cost of the bill if a drink costs $\$3$, a meal costs $\$20$, an appetizer costs $\$5.50$, and a slice of cake costs $\$3.75$?**

$12d + 12m + 3a + 6c$

$12(3) + 12(20) + 3(5.50) + 6(3.75)$

$36 + 240 + 16.50 + 22.50$

315

The cost of the bill is $\$315$.

d. **Suppose the family had a 10% discount coupon for the entire check and then left an 18% tip. What is the total?**

$(315 - 315(0.10)) + 0.18(315 - 315(0.10))$

$1.18(315 - 315(0.10))$

$1.18(315(0.90))$

334.53

After the discount and tip, the new total is $\$334.53$.

2. **Sally designs web pages for customers. She charges $\$135.50$ per web page; however, she must pay a monthly rental fee of $\$650$ for her office. Write an expression to determine her take-home pay after expenses. If Sally designed 5 web pages last month, what was her take-home pay after expenses?**

$w =$ *number of webpages Sally designs*

$135.50w - 650$

$135.50(5) - 650$

27.50

After expenses, Sally's take-home pay is $\$27.50$.

3. **While shopping, Megan and her friend Rylie find a pair of boots on sale for 25% off the original price. Megan calculates the final cost of the boots by first deducting the 25% and then adding the 6% sales tax. Rylie thinks Megan will pay less if she pays the 6% sales tax first and then takes the 25% discount.**

a. **Write an expression to represent each girl's scenario if the original price of the boots was x dollars.**

Megan	*Rylie*
$(x - 0.25x) + 0.06(x - 0.25x)$	$(x + 0.06x) - 0.25(x + 0.06x)$
$1.06(x - 0.25x)$	$0.75(x + 0.06x)$
$1.06(0.75x)$	$0.75(1.06x)$

b. **Evaluate each expression if the boots originally cost $\$200$.**

Megan	*Rylie*
$1.06(0.75x)$	$0.75(1.06x)$
$1.06(0.75(200))$	$0.75(1.06(200))$
159	159

Using both Megan's and Rylie's methods would show that the boots would cost $\$159$.

c. **Who was right? Explain how you know.**

Neither girl was right. They both pay the same amount.

d. **Explain how both girls' expressions are equivalent.**

Two expressions are equivalent if they yield the same number for every substitution of numbers for the variables in each expression. Since multiplication is commutative, the order of the multiplication can be reversed, and the result will remain the same.

Lesson 20: Investments—Performing Operations with Rational Numbers

Student Outcomes

- Students perform various calculations involving rational numbers to solve a problem related to the change in an investment's balance over time.
- Students recognize and use mathematics as a tool to solve real-life problems.

Classwork

Mathematical Modeling Exercise (25 minutes): College Investments

Students are given records for an investment fund over the past 5 years. The records include beginning balance, semiannual statements, and additional fees. Students have a 4-part task:

1. Determine the balance at the end of 5 years.
2. Determine the annual gain or loss and the overall 5-year gain or loss.
3. Analyze the result and write a comparative conclusion, defending or refuting their conclusion.
4. Answer questions related to the investment data.

Familiarize students with the format of the register. Discuss how to complete the register by completing the first six months as a class together. Then have students individually complete the rest of the register for the remaining time.

Suggestion: Allow students to use calculators to assist in the arithmetic.

Mathematical Modeling Exercise: College Investments

Justin and Adrienne deposited $\$20,000$ into an investment account for 5 years. They hoped the money invested and the money made on their investment would amount to at least $\$30,000$ to help pay for their daughter's college tuition and expenses. The account they chose has several benefits and fees associated with it. Every 6 months, a summary statement is sent to Justin and Adrienne. The statement includes the amount of money either gained or lost. Below are semiannual (twice a year) statements for a period of 5 years. In addition to the statements, the following information is needed to complete the task:

- **For every statement, there is an administrative fee of $\$15$ to cover costs such as secretarial work, office supplies, and postage.**
- **If there is a withdrawal made, a broker's fee is deducted from the account. The amount of the broker's fee is 2% of the transaction amount.**

TASK: Using the above information, semiannual statements, register, and beginning balance, do the following:

1. **Record the beginning balance and all transactions from the account statements into the register.**
2. **Determine the annual gain or loss as well as the overall 5-year gain or loss.**
3. **Determine if there is enough money in the account after 5 years to cover $\$30,000$ of college expenses for Justin and Adrienne's daughter. Write a summary to defend your answer. Be sure to indicate how much money is in excess, or the shortage that exists.**
4. **Answer the related questions that follow.**

(Note: This activity may be adapted for use with spreadsheet software.)

College Investment Fund Semi-Annual Statement

January 1, 2008 – June 30, 2008

Investment Gain/(Loss) : 700.00

College Investment Fund Semi-Annual Statement

July 1, 2008 – December 31, 2008

Investment Gain/(Loss): 754.38

College Investment Fund Semi-Annual Statement

January 1, 2009 – June 30, 2009

Investment Gain/(Loss): (49.88)

College Investment Fund Semi-Annual Statement

July 1, 2009 – December 31, 2009

Withdrawal: 500.00

Investment Gain/(Loss: (17.41)

College Investment Fund Semi-Annual Statement

January 1, 2010 – June 30, 2010

Investment Gain/(Loss): 676.93

College Investment Fund Semi-Annual Statement

July 1, 2010 – December 31, 2010

Investment Gain/(Loss): 759.45

College Investment Fund Semi-Annual Statement

January 1, 2011 – June 30, 2011

Deposit: 1,500.00

Investment Gain/(Loss): 880.09

College Investment Fund Semi-Annual Statement

July 1, 2011 – December 31, 2011

Investment Gain/(Loss) : 922.99

College Investment Fund Semi-Annual Statement

January 1, 2012 – June 30, 2012

Deposit: 800.00

Investment Gain/(Loss): 942.33

College Investment Fund Semi-Annual Statement

July 1, 2012 – December 31, 2012

Investment Gain/(Loss): 909.71

5. **Register**

DATE	DESCRIPTION OF TRANSACTION	WITHDRAWAL	DEPOSIT	BALANCE	EXPRESSION
	Beginning Balance	---	---	$20,000.00	$20,000.00
Jan. – June: 2008	*Gain on investment*		700.00	20,700.00	20,000 + 700
	Administrative fee	15.00		20,685.00	20,700 − 15
July – Dec.: 2008	*Gain on investment*		754.38	21,439.38	20,685 + 754.38
	Administrative fee	15.00		21,424.38	21,439.38 − 15
Jan. – June: 2009	*Loss on investment*	49.88		21,374.50	21,424.38 − 49.88
	Administrative fee	15.00		21,359.50	21,374.50 − 15
July – Dec.: 2009	*Withdrawal*	500.00		20,859.50	21,359.50 − 500
	Broker fee	10.00		20,849.50	20,859.50 − 10
	Loss on investment	17.41		20,832.09	20,849.50 − 17.41
	Administrative fee	15.00		20,817.09	20,832.09 − 15
Jan. – June: 2010	*Gain on investment*		676.93	21,494.02	20,817.09 + 676.93
	Administrative fee	15.00		21,479.02	21,494.02 − 15
July – Dec.: 2010	*Gain on investment*		759.45	22,238.47	21,479.02 + 759.45
	Administrative fee	15.00		22,223.47	22,238.47 − 15
Jan. – June: 2011	*Deposit*		1,500.00	23,723.47	22,223.47 + 1500
	Gain on investment		880.09	24,603.56	23,723.47 + 880.09
	Administrative fee	15.00		24,588.56	24,603.56 − 15
July – Dec.: 2011	*Gain on investment*		922.99	25,511.55	24,588.56 + 922.99
	Administrative fee	15.00		25,496.55	25,511.55 − 15
Jan. – June: 2012	*Deposit*		800.00	26,296.55	25,496.55 + 800
	Gain on investment		942.33	27,238.88	26,296.55 + 942.33
	Administrative fee	15.00		27,223.88	27,238.88 − 15
July – Dec.: 2012	*Gain on investment*		909.71	28,133.59	27,223.88 + 909.71
	Administrative fee	15.00		28,118.59	28,133.59 − 15
Prediction:					
Jan. – June: 2013	*Gain on investment*		900.00	29,018.59	28,118.59 + 900
	Administrative fee	15.00		29,003.59	29,018.59 − 15

- Describe the process of completing the register.
  - *Starting with the beginning balance, fill in the description of the transaction and the amount. If the transaction is an investment loss, withdrawal, or fee, then the amount is recorded in the payment column. If the transaction is an investment gain or deposit, then the amount is recorded in the deposit column. To obtain the new balance, subtract the payment amount, or add the deposit amount, from the balance on the preceding line. Record the new balance, and use that balance to complete the next line.*

> *Scaffolding:*
>
> Discuss what a register is and how it is used to organize a series of transactions. Also, discuss how a loss can be represented by using parentheses (e.g., (607.29)).

- Describe how to find the broker's fee.
  - *The broker's fee is* 2% *of the transaction amount. To find the broker's fee, you must first find the total of the transaction amount. Once you have that, write the percent as a fraction out of* 100 *and multiply the fraction by the transaction amount. This result is the amount of the broker's fee, which is then subtracted from the preceding balance.*

    Example: 2% *of* $\$2{,}500$

    $$\frac{2}{100} \times 2{,}500 = \frac{1}{50} \times 2{,}500 = 50$$

- Compare your register with the person next to you. Did each of you list the transactions in the same order? Does it make a difference?
  - *The order is probably not the same. The order of the transactions for each* 6*-month period does not make a difference.*
- Continue to compare your registers. Do you both get the same balance at the end of 2012? If not, switch papers and check to see if you can find your neighbor's mistake.

6. **Annual Gain/Loss Summary**

Year	Total Gain/(Loss)	Numerical Expression
2008	$1,424.38$	$21,424.38 - 20,000$
2009	(607.29)	$20,817.09 - 21,424.38$
2010	$1,406.38$	$22,223.47 - 20,817.09$
2011	$3,273.08$	$25,496.55 - 22,223.47$
2012	$2,622.04$	$28,118.59 - 25,496.55$
5-Year Gain/Loss	$8,118.59$	$28,118.59 - 20,000$

- How do you find the annual gain or loss for each year?
  - *Subtract the ending balance from the beginning balance. If the beginning balance is smaller than the ending balance, then the difference is a gain. If the beginning balance is larger than the ending balance, then the difference is a loss and written within ( ).*
- In what year was the greatest gain? In what year was the greatest loss?
  - *The greatest gain was in 2011. The greatest loss was in 2009 because of the withdrawal of* $\$500$ *and the losses on the investment.*
- How does knowing the overall gain or loss assist in writing the comparative conclusion?
  - *Once you know the gain or loss, you can find the ending balance. Since the ending balance was not at least* $\$30{,}000$*, there was not enough money to cover the college expenses.*
- Using the 5-year total gain or loss figures, write an expression using positive and negative rational numbers that can be solved to find the total gain or loss.
  - $1{,}424.38 + (-607.29) + 1{,}406.38 + 3{,}273.08 + 2{,}622.04$
  - $-607.29 + 1{,}424.38 + 1{,}406.38 + 3{,}273.08 + 2{,}622.04$

> *Scaffolding:*
>
> Review or reiterate that the operation associated with payments is subtraction, and the operation associated with deposits is addition.

7. **Summary**

   There is not enough money in the account at the end of 5 years to cover the college expenses, but it is close. They needed at least $\$30,000$ in the account to cover the expenses, and there was $\$28,118.59$, leaving a shortage of $\$1,881.41$.

8. **Related Questions**

   a. **For the first half of 2009, there was a $\$700$ gain on the initial investment of $\$20,000$. Represent the gain as a percentage of the initial investment.**

   $\frac{x}{100} = \frac{700}{20{,}000}$ *The gain was 3.5% of $\$20,000$.*

   b. **Based on the gains and losses on their investment during this 5-year period, over what period of time was their investment not doing well? How do you know? What factors might contribute to this?**

   The investment was not doing well in 2009. There were losses on the investment for both halves of the year, and $\$500$ was taken out of the account. It could be because the economy was doing badly, and a recession affected the investment's performance.

   c. **In math class, Jaheim and Frank were working on finding the total amount of the investment after 5 years. As a final step, Jaheim subtracted $\$150$ for administrative fees from the balance he arrived at after adding in all the deposits and subtracting out the one withdrawal and broker's fee. For every semiannual statement, Frank subtracted $\$15$ from the account balance for the administrative fee. Both boys arrived at the same ending 5-year balance. How is this possible? Explain.**

   Jaheim took the $\$15$ fee and multiplied it by 10, since there were 10 statements, and deducted the $\$150$ total. Frank subtracted $\$15$ from the account balance for each statement. That was 10 times. So, both ways produce the same result: reducing the account balance by $\$150$ overall.

   d. **Based on the past statements for their investment account, predict what activity you might expect to see on Adrienne and Justin's January–June 2013 account statement. Then record it in the register to arrive at the balance as of June 30, 2013.**

   I predict the account will continue to produce gains. The gains have been around $\$900$ for the past four statements, so I predict it will be about $\$900$ again, since it decreased by a little bit the last time, and there was a $\$909.71$ gain the last time. If I take away $\$15$ for the administrative fee, the balance would go up by $\$885$, and it would be $\$29,003.59$.

   e. **Using the answer from part (d), if their daughter's college bill is due in September of 2013, how much money do you estimate will be in their investment account at the end of August 2013 before the college bill is paid? Support your answer.**

   Their investment could gain more money for July and August. Right now, it is gaining about $\$900$ per statement. If I divide that by 6, it equals $\$150$ (which is the average gain per month). So, for July and August I estimate that it will earn about another $\$300$ (including the $\$15$ fee), so there might be $\$29,333.59$ in the account.

Exercise (10 minutes)

MP.2 Students are given a transaction log of a business entertainment account. The transactions are completed for students and the ending balance is given as well. Students are required to work *backward* to find the beginning balance.

Exercise

Below is a transaction log of a business entertainment account. The transactions are completed, and the ending balance in the account is $\$525.55$. Determine the beginning balance.

DATE	DESCRIPTION OF TRANSACTION	PAYMENT	DEPOSIT	BALANCE
	Beginning Balance	---	---	801.02
12/1/10	Bargain Electronic (i-Pod)	199.99		601.03
12/5/10	Lenny's Drive-Up (Gift Certificate)	75.00		526.03
12/7/10	Check from Customer: Reynolds		200.00	726.03
12/15/10	Pasta House (Dinner)	285.00		441.03
12/20/10	Refund from Clear's Play House		150.00	591.03
12/22/10	Gaffney's Tree Nursery	65.48		525.55

- When the beginning balance was given, the transactions were subtracted from the balance if a payment was made, and the deposits were added to the balance if a deposit was made. How does that process change when the ending balance is given and the challenge is to find the beginning balance?
    - *If the ending balance is given and a payment was made, then you need to add the payment to the ending balance to get the beginning balance. Likewise, if the ending balance is given and a deposit was made, then you need to subtract the payment from the ending balance to get the beginning balance.*
- Model the process described in the previous question by writing and solving an equation for the deposit made from the refund from Clear's Play House. Assume the preceding balance was x.
    - *Refund from Clear's Play House:*

$$x + 150 = 591.03$$
$$x + 150 - 150 = 591.03 - 150$$
$$x = 441.03$$

- What happens if the deposit amount is greater than the ending balance? How can this be written?
    - *If the deposit is greater than the ending balance, then the beginning balance would be less than 0 and written as a negative number. This negative number indicates owing money.*

Closing (5 minutes)

Answer any remaining questions.

- What role do rational numbers play in solving real-world problems?
    - *Answers will vary, but students will most likely talk about recording investments and transactions.*

Lesson Summary

- Calculations with rational numbers are used when recording investment transactions.
- Deposits are added to an account balance; money is deposited into the account.
- Gains are added to an account balance; they are positive returns on the investment.
- Withdrawals are subtracted from an account balance; money is taken out of the account.
- Losses are subtracted from an account balance; they are negative returns on the investment.
- Fees are subtracted from an account balance; the bank or financial company is charging you for a service.

Exit Ticket (5 minutes)

Name ______________________________ Date ______________

Lesson 20: Investments—Performing Operations with Rational Numbers

Exit Ticket

1. Using the incomplete register below, work forward and backward to determine the beginning and ending balances after the series of transactions listed.

DATE	DESCRIPTION OF TRANSACTION	PAYMENT	DEPOSIT	BALANCE
	Beginning Balance	---	---	
1/31/12	Paycheck		350.55	
2/1/2012	Gillian's Chocolate Factory (Candy)	32.40		685.26
2/4/12	Main Street Jeweler's	425.30		
2/14/12	Saratoga Steakhouse	125.31		

2. Write an expression to represent the balance after the paycheck was deposited on 1/31/12. Let x represent the beginning balance.

3. Write a numerical expression to represent the balance after the transaction for Main Street Jeweler's was made.

Exit Ticket Sample Solutions

1. Using the incomplete register below, work forward and backward to determine the beginning and ending balances after the series of transactions listed.

DATE	DESCRIPTION OF TRANSACTION	PAYMENT	DEPOSIT	BALANCE
	Beginning Balance	---	---	367.11
1/31/12	Paycheck		350.55	717.66
2/1/12	Gillian's Chocolate Factory (Candy)	32.40		685.26
2/4/12	Main Street Jeweler's	425.30		259.96
2/14/12	Saratoga Steakhouse	125.31		134.65

2. Write an expression to represent the balance after the paycheck was deposited on 1/31/12. Let x represent the beginning balance.

   $x + 350.55$

3. Write a numerical expression to represent the balance after the transaction for Main Street Jeweler's was made.

   $685.26 - 425.30$

Problem Set Sample Solutions

1. You are planning a fundraiser for your student council. The fundraiser is a Glow in the Dark Dance. Solve each entry below, and complete the transaction log to determine the ending balance in the student account.

   a. The cost of admission to the dance is $\$7$ per person, and all tickets were sold on November 1. Write an expression to represent the total amount of money collected for admission. Evaluate the expression if 250 people attended the dance.

   p represents the number of people attending dance.

   $7p$

   $7(250) = 1750$

   If 250 people attended the dance, $\$1,750$ would be collected.

   b. The following expenses were necessary for the dance, and checks were written to each company.

   - DJ for the dance—*Music Madness DJ* costs $\$200$ and paid for on November 3.
   - Glow sticks from *Glow World, Inc.* for the first 100 entrants. Cost of glow sticks were $\$0.75$ each plus 8% sales tax and bought on November 4.

   Complete the transaction log below based on this information

DATE	DESCRIPTION OF TRANSACTION	PAYMENT	DEPOSIT	BALANCE
	Beginning Balance	---	---	$1,243.56$
November 1	*Dance Admission*		$1,750.00$	$2,993.56$
November 3	*Music Madness DJ*	200.00		$2,793.56$
November 4	*Glow Sticks from Glow World, Inc.*	81.00		$2,712.56$

c. Write a numerical expression to determine the cost of the glow sticks.

$100\left(\frac{8}{100}(0.75) + 0.75\right)$

All 100 *glow sticks cost* $\$81$.

Analyze the results.

d. Write an algebraic expression to represent the profit earned from the fundraiser. (Profit is the amount of money collected in admissions minus all expenses.)

$7p - 200 - 81$

$7p + (-200) + (-81)$

$7p + (-281)$ *or* $7p - 281$

e. Evaluate the expression to determine the profit if 250 people attended the dance. Use the variable p to represent the number of people attending the dance (from part (a)).

$7p + (-281)$

$7(250) + (-281)$

$1,750 + (-281)$

$1,469$

The profit is $\$1,469$.

f. Using the transaction log above, what was the amount of the profit earned?

$2,712.56 - 1,243.56 = 1469$ *The profit is* $\$1,469$.

2. The register below shows a series of transactions made to an investment account. Vinnie and Anthony both completed the register in hopes of finding the beginning balance. As you can see, they do not get the same answer. Who was correct? What mistake did the other person make? What was the monthly gain or loss?

Original Register

DATE	DESCRIPTION OF TRANSACTION	PAYMENT	DEPOSIT	BALANCE
	Beginning Balance	---	---	
3/1/11	Broker's Fee	250.00		
3/10/11	Loan Withdrawal	895.22		
3/15/11	Refund – Misc. Fee		50.00	
3/31/11	Investment Results		2,012.22	18,917.00

EUREKA MATH™

Vinnie's Work

DATE	DESCRIPTION OF TRANSACTION	PAYMENT	DEPOSIT	BALANCE
	Beginning Balance	---	---	$18,000.00$
3/1/11	Broker's Fee	250.00		$17,750.00$
3/10/11	Loan Withdrawal	895.22		$16,854.78$
3/15/11	Refund – Misc. Fee		50.00	$16,904.78$
3/31/11	Investment Results		$2,012.22$	$18,917.00$

Anthony's Work

DATE	DESCRIPTION OF TRANSACTION	PAYMENT	DEPOSIT	BALANCE
	Beginning Balance	---	---	$19,834.00$
3/1/11	Broker's Fee	250.00		$20,084.00$
3/10/11	Loan Withdrawal	895.22		$20,979.22$
3/15/11	Refund – Misc. Fee		50.00	$20,929.22$
3/31/11	Investment Results		$2,012.22$	$18,917.00$

The correct register is Vinnie's.

Anthony made the mistake of using the operations for moving forward. He added the deposits and subtracted the payments, but since he was working backward in the problem, he needed to do just the opposite.

The monthly gain was $\$917$. *This was a gain because the ending balance was greater than the beginning balance, and the amount of the gain was calculated by* $18,917 - 18,000 = 917$.

Lesson 21: If–Then Moves with Integer Number Cards

Student Outcomes

- Students understand that if a number sentence is true, and any of the following changes are made to the number sentence, the resulting number sentence will be true:
  - i. Adding the same number to both sides of the equation

    If $a = b$, then $a + c = b + c$.
  - ii. Subtracting the same number from both sides of the equation

    If $a = b$, then $a - c = b - c$.
  - iii. Multiplying each side of the equation by the same number

    If $a = b$, then $a(c) = b(c)$.
  - iv. Dividing each side of the equation by the same nonzero number

    If $a = b$ and $c \neq 0$, then $a \div c = b \div c$.
- Students revisit the Integer Game to justify the if–then statements referenced above.

Classwork

Exploratory Challenge (25 minutes): Integer Game Revisited

Pass out three integer number cards to each student, using integers from -2 to 2. Have students, on their student pages, record their cards and their total score (sum). The scores will be between -6 and 6, inclusive. If there are more than 13 students, at least two will have the same score.

Have students find a classmate with the same score, and have them sit next to each other. Students with tied scores should compare their initial cards, noting they are probably different cards with the same sum.

MP.2

Select a pair of students with equal sums, and have them write their cards and scores on the board. Continue playing the game with the following changes. Have students, in their student materials, describe the event, record their new sums, and write overall conclusions using if–then statements based on each event of the game.

Exploratory Challenge: Integer Game Revisited

Let's investigate what happens if a card is added or removed from a hand of integers.

My cards:

My score:

Event 1

MP.2

Give each pair of students two more integer cards (one for each student) containing the same positive value, and ask them to record the change and the resulting score. (For instance, a 3 card is given to each partner, both of whom had a previous card total of -1, and both students determine that their card totals remain equal, as now they each have a score of 2.)

Event 1

My new score:

Conclusion:

Possible solution:

	Partner 1			Partner 2		
Original cards:	-1	2	-2	0	1	-2
Original score:		-1			-1	
Event 1 (both partners receive the card 2)						
New score:		1			1	

Repeat this process with one minor change; this time both students receive one integer card containing the same negative value. Have students record their new scores and, after comparing with their partners, write a conclusion using an if–then statement.

Event 1 (both partners receive the card -1)		
New score:	0	0
Conclusion:	*If the sums are equal, then the same negative or positive number added to the sums will remain equal.*	

Series of questions leading to the conclusion:

- Were your scores the same when we began?
  - *Yes*
- Did you add the same values to your hand each time?
  - *Yes*
- Did the value of your hand change each time you added a new card?
  - *Yes*
- Was the value of your hand still the same as your partner's after each card was added?
  - *Yes*

- Why did the value of your hand remain the same after you added the new cards?
  - *We started with the same sum; therefore, when we added a new card, we had equivalent expressions, which resulted in the same sum.*
- Since your original cards were different, but your original sum was the same, write a conclusion that was exemplified by this event.
  - *If the original sums were equal, you can add a number, either positive or negative, and the sums will remain equal.*

Event 2

Pick either the same pair of students or another pair who have original sums that are equal *and* have at least one identical card. If possible, pick two groups to go to the board. One group will have an identical positive card; the other will have an identical negative card.

Event 2

My new score:

Conclusion:

If there are two students without the same scores, then use the example that follows.

Student 1: $-2, -1, 2$

Student 2: $0, -2, 1$

Instruct students to remove the identical card from their partner's hand and record their new score. In the student materials, students are asked to describe the event, record their new scores, compare their scores to their partners', write numerical expressions based on the cards, and write overall conclusions based on the event.

- Compare each of your cards to your partner's. Do you have the exact same two cards remaining?
  - *Probably not*
- Compare your new sum to your partner's new sum. What happened?
  - *The sums stayed the same.*
- Write a conclusion that explains what happens when the sums of your cards remain the same when the same card is removed.
  - *If the original sums were equal, you can subtract a number, either positive or negative, and the sums will remain equal.*

Sample solution:

	Partner 1	*Partner 2*
	$-2, -1, 2$	$0, -2, 1$
Score:	-1	-1
Remove identical cards; remove -2.		
New score:	1	1
Numerical expression:	$-2 + -1 + 2 - (-2)$	$0 + -2 + 1 - (-2)$

Conclusion: *If the original sums are equal, you can subtract the same number, either positive or negative, and the sums will remain equal.*

Event 3

Instruct students to look at their original three cards. Double or triple (if there are enough cards) each student's cards with cards matching their original cards. In the student materials, students are asked to describe the event, write the sum as a numerical expression, record the new score, compare it to their partner's, and write an overall conclusion based on the event.

Event 3

My new score:

Expression:

Conclusion:

Possible solution:

Original cards:	$1, \quad 2, \quad 2$		
Score:	5		
Triple the cards.			
New score:	15		
Numerical expression:	$1+2+2+1+2+2+1+2+2$	*or*	$3(1+2+2)$
Conclusion:	15		15

If the sums are the same, then the product of the sums will remain equal when both are multiplied by the same rational number.

- Compare your original sum to your new sum. What happened?
  - *It is doubled or tripled (if enough cards).*
- Compare your new sum to your partner's new sum. What happened?
  - *They are the same.*
- Look at your numerical expression to find the sum. For students who used only addition or repeated addition, look to see how you could have multiplied. For students who multiplied, what property is applied to get the solution?
  - *Repeated addition could be written as multiplication. The distributive property is then applied to simplify the expression.*
- Write a conclusion about the effects of multiplying a sum by a number.
  - *If the sums of two sets of numbers are equal, then when those numbers are multiplied by another number, the sums will be multiplied by the same number and remain equal.*

Event 4

Select a pair of students with sums of either 4 or -4 to come to the board. Now give them both integer cards with the same nonzero value. Instruct students to divide the original sum by the new card. In their student materials, students are to describe the event, write a numerical expression, and write a conclusion based on the results shown at the front of the class.

- Compare your cards to your partner's. What can you conclude about your original cards and sum?
  - *Original cards are probably different, but the sums are the same.*
- Compared to your partner's, what happened to the sum when you divided by the same integer card?
  - *The sums are different from the original sums but remain equal to each other.*
- Write a conclusion that describes the effects of dividing equal sums by an identical number.
  - *If the sums are the same, then the quotient of the sums will remain equal when both are divided by the same rational number.*

> *Scaffolding:*
>
> This is an additional option for teachers with proficient students.
>
> Instruct one person from the pair to put together as many cards as possible so that the sum of the numbers on the cards is between -2 and 2. Have students make the following trade: If one person has a card equal to the value of the new sum, then trade the one card whose value is the sum for ALL of the other cards giving that sum. Calculate the new sum of remaining original cards with ALL of the new cards. In the student materials, students are to describe the event and summarize the results.

Event 4

Expression:

Conclusion:

Possible solution:

	Partner 1	*Partner 2*
Original cards:	$2,\ 2,\ 0$	$2,\ 1,\ 1$
Score:	4	4
Given card value:	-2	-2
Quotient:	-2	-2
Numerical expression:	$\frac{(2+2+0)}{-2}$	$\frac{(2+1+1)}{-2}$

Conclusion: *If the sums are the same, then the quotient of the sums will remain equal when both are divided by the same rational number.*

Discussion

Discuss the overall conclusions that if two quantities are equal, then you can add, subtract, multiply, or divide a number to both quantities, and the resulting quantities will be equal.

- Explain why the sum remains the same if you received many more cards.
  - *The cards you received in total were equal to the card you traded. You may have received many more cards, but the overall sum did not change because what you gave away was the same as what you gained.*

Exercises (10 minutes)

Have students complete the first row of the table individually and then compare their results with a partner.

Exercises

1. **The table below shows two hands from the Integer Game and a series of changes that occurred to each hand. Part of the table is completed for you. Complete the remaining part of the table; then summarize the results.**

	Hand 1	Result	Hand 2	Result
Original	$1+(-4)+2$	-1	$0+5+(-6)$	-1
Add 4	$(1+(-4)+2)+4$	3	$(0+5+(-6))+4$	3
Subtract 1	$((1+(-4)+2)+4)-1$	2	$((0+5+(-6))+4)-1$	2
Multiply by 3	$3(((1+(-4)+2)+4)-1)$	6	$3(((0+5+(-6))+4)-1)$	6
Divide by 2	$(3(((1+(-4)+2)+4)-1))\div 2$	3	$(3(((0+5+(-6))+4)-1))\div 2$	3

Since the sums of each original hand are the same, the same cards can be added, subtracted, multiplied, and divided, and the sums will remain equal to each other.

Perform each of the indicated operations to each expression, compare the new results, and write a conclusion.

- Does it matter if you perform the operation to the original numerical expression or to the original answer?
  - *It does not matter. Doing it both ways would be a good check.*

2. **Complete the table below using the multiplication property of equality.**

	Original expression and result	Equivalent expression and result
	$3+(-5)=-2$	$-4+2=-2$
Multiply both expressions by -3	$-3(3+-5)=-3(-2)=6$	$-3(-4+2)=-3(-2)=6$
Write a conclusion using if–then	*If* $3+-5=-4+2$, *then* $-3(3+5)=-3(-4+2)$	

Closing (2 minutes)

Describe additional questions.

- While playing the Integer Game, you and your partner each add a card with the same value to your hand. After doing this, you and your partner have the same score. How is this possible?
  - *This is only possible if we started with equivalent sums.*
- While playing the Integer Game, you and your partner have equal scores before and after removing a card from each of your hands. How is this possible?
  - *This is only possible if both partners removed the same card from their hand.*

Lesson Summary

- If a number sentence is true, and the same number is added to both sides of the equation, then the resulting number sentence is true. ***(addition property of equality)***
- If a number sentence is true, and the same number is subtracted from both sides of the equation, then the resulting number sentence is true. ***(subtraction property of equality)***
- If a number sentence is true, and both sides of the equation are multiplied by the same number, then the resulting number sentence is true. ***(multiplication property of equality)***
- If a number sentence is true, and both sides of the equation are divided by the same nonzero number, then the resulting number sentence is true. ***(division property of equality)***

Exit Ticket (8 minutes)

Name ______________________________ Date ______________

Lesson 21: If–Then Moves with Integer Number Cards

Exit Ticket

Compare the two expressions: Expression 1: $6 + 7 + -5$

Expression 2: $-5 + 10 + 3$

1. Are the two expressions equivalent? How do you know?

2. Subtract -5 from each expression. Write the new numerical expression, and write a conclusion as an if–then statement.

3. Add 4 to each expression. Write the new numerical expression, and write a conclusion as an if–then statement.

4. Divide each expression by -2. Write the new numerical expression, and write a conclusion as an if–then statement.

Exit Ticket Sample Solutions

Compare the two expressions. Expression 1: $6 + 7 + -5$

Expression 2: $-5 + 10 + 3$

1. Are the two expressions equivalent? How do you know?

   Yes, the expressions are equivalent because Expression 1 is equal to 8 and Expression 2 is equal to 8, as well. When two expressions evaluate to the same number, they are equivalent.

2. Subtract -5 from each expression. Write the new numerical expression, and write a conclusion as an if–then statement.

   Expression 1: $6 + 7 \pm 5 - (-5)$ *Expression 2:* $-5 + 10 + 3 - (-5)$

   13 13

   If $6 + 7 + -5 = -5 + 10 + 3$, *then* $6 + 7 + -5 - (-5) = -5 + 10 + 3 - (-5)$.

   If Expression 1= Expression 2, then (Expression 1$-(-5)$*) = (Expression 2*$-(-5)$*).*

3. Add 4 to each expression. Write the new numerical expression, and write a conclusion as an if–then statement.

   Expression 1: $6 + 7 \pm 5 + 4$ *Expression 2:* $-5 + 10 + 3 + 4$

   12 12

   If $6 + 7 + -5 = -5 + 10 + 3$, *then* $6 + 7 + -5 + 4 = -5 + 10 + 3 + 4$.

   If Expression 1 = Expression 2, then (Expression 1 $+ 4$*) = (Expression 2* $+ 4$*).*

4. Divide each expression by -2. Write the new numerical expression, and write a conclusion as an if–then statement.

   Expression 1: $(6 + 7 + -5) \div -2$ *Expression 2:* $(-5 + 10 + 3) \div -2$

   $8 \div -2$ $8 \div -2$

   -4 -4

   If $6 + 7 + -5 = -5 + 10 + 3$, *then* $(6 + 7 + -5) \div -2 = (-5 + 10 + 3) \div -2$

   If Expression 1= Expression 2, then (Expression 1 $\div -2$*) = (Expression 2* $\div -2$*).*

Problem Set Sample Solutions

This Problem Set provides students with additional practice evaluating numerical expressions and applying different moves while seeing the effect on number sentences.

1. Evaluate the following numerical expressions.

   a. $2 + (-3) + 7 = 6$

   b. $-4 - 1 = -5$

   c. $-\frac{5}{2} \times 2 = -5$

   d. $-10 \div 2 + 3 = -2$

   e. $\left(\frac{1}{2}\right)(8) + 2 = 6$

   f. $3 + (-4) - 1 = -2$

2. **Which expressions from Exercise 1 are equal?**

Expressions (a) and (e) are equivalent.

Expressions (b) and (c) are equivalent.

Expressions (d) and (f) are equivalent.

3. **If two of the equivalent expressions from Exercise 1 are divided by 3, write an if–then statement using the properties of equality.**

If $2 + (-3) + 7 = \left(\frac{1}{2}\right)(8) + 2$, *then* $(2 + (-3) + 7) \div 3 = \left(\left(\frac{1}{2}\right)(8) + 2\right) \div 3$.

4. **Write an if–then statement if -3 is multiplied by the following equation: $-1 - 3 = -4$.**

If $-1 - 3 = -4$, *then* $-3(-1 - 3) = -3(-4)$

5. **Simplify the expression.**

$5 + 6 - 5 + 4 + 7 - 3 + 6 - 3$

$= 17$

Using the expression, write an equation.

$5 + 6 - 5 + 4 + 7 - 3 + 6 - 3 = 17$

Rewrite the equation if 5 is added to both expressions.

$5 + 6 - 5 + 4 + 7 - 3 + 6 - 3 + 5 = 17 + 5$

Write an if–then statement using the properties of equality.

If $5 + 6 - 5 + 4 + 7 - 3 + 6 - 3 = 17$, *then* $5 + 6 - 5 + 4 + 7 - 3 + 6 - 3 + 5 = 17 + 5$

Lesson 22: Solving Equations Using Algebra

Student Outcomes

- Students use algebra to solve equations (of the form $px + q = r$ and $p(x + q) = r$, where p, q, and r are specific rational numbers), using techniques of making zero (adding the additive inverse) and making one (multiplying by the multiplicative inverse) to solve for the variable.
- Students identify and compare the sequence of operations used to find the solution to an equation algebraically, with the sequence of operations used to solve the equation with tape diagrams. They recognize the steps as being the same.
- Students solve equations for the value of the variable using inverse operations, by making zero (adding the additive inverse) and making one (multiplying by the multiplicative inverse).

Classwork

In this lesson, you will transition from solving equations using tape diagrams to solving equations algebraically by *making zero* (using the additive inverse) and *making one* (using the multiplicative inverse). Justify your work by identifying which algebraic property you used for each step in solving the problems. Explain your work by writing out how you solved the equations step by step and relate each step to those used with a tape diagram.

Example 1 (10 minutes): Yoshiro's New Puppy

MP.1 Use this problem to emphasize the use of illustrating the problem and solving an algebraic problem with a tape diagram. Drawing the puppy yard will help show students the meaning of perimeter and make sense of the problem.

Example 1: Yoshiro's New Puppy

Yoshiro has a new puppy. She decides to create an enclosure for her puppy in her backyard. The enclosure is in the shape of a hexagon (six-sided polygon) with one pair of opposite sides running the same distance along the length of two parallel flower beds. There are two boundaries at one end of the flower beds that are 10 ft. and 12 ft., respectively, and at the other end, the two boundaries are 15 ft. and 20 ft., respectively. If the perimeter of the enclosure is 137 ft., what is the length of each side that runs along the flower bed?

- What is the general shape of the puppy yard? Draw a sketch of the puppy yard.

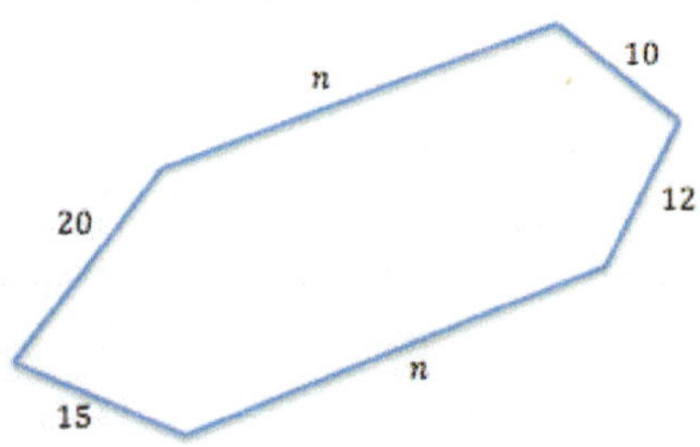

Scaffolding:

Have students write out in words what they will do to transition from words to algebraic symbols.

MP.4

- Write an equation that would model finding the perimeter of the puppy yard.
  - *The sum of the lengths of the sides* $=$ *Perimeter*

    $n + n + 10 + 12 + 20 + 15 = 137$

- Model and solve this equation with a tape diagram.
  - *Sample response:*

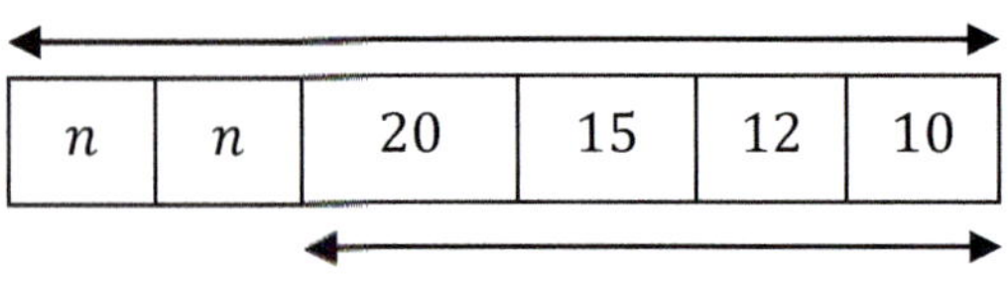

$137 - 57 = 80;\ 80 \div 2 = 40$

Now review *making zero* in an equation and *making one* in an equation. Explicitly connect *making zero* and *making one* in the next question to the bar model diagram. Subtracting 57 from 137 in the bar diagram is the same as using the subtraction property of equality (i.e., subtracting 57 from both sides of the equation in order to make zero). Dividing 80 by 2 to find the size of two equal groups that total 80 is the same as using the multiplicative property of equality (i.e., multiplying each side of the equation by $\frac{1}{2}$ to make one group of n).

- Use algebra to solve this equation.
  - *First, use the additive inverse to find out what the lengths of the two missing sides are together. Then, use the multiplicative inverse to find the length of one of the two equal sides. Sum of missing sides + Sum of known sides = Perimeter*

If:	$2n + 57 = 137$	
Then:	$2n + 57 - 57 = 137 - 57$	*Subtraction property of equality*
If:	$2n + 0 = 80$	
Then:	$2n = 80$	*Additive identity*
If:	$2n = 80$	
Then:	$\frac{1}{2}(2n) = \frac{1}{2}(80)$	*Multiplication property of equality*
If:	$1n = 40$	
Then:	$n = 40$	*Multiplicative identity*

- Does your solution make sense in this context? Why?
  - *Yes,* 40 ft. *makes sense because when you replace the two missing sides of the hexagon with* 40 *in the number sentence (* $40 + 40 + 10 + 12 + 20 + 15 = 137$*), the lengths of the sides reach a total of* 137.

Example 2 (10 minutes): Swim Practice

> **Example 2: Swim Practice**
>
> **Jenny is on the local swim team for the summer and has swim practice four days per week. The schedule is the same each day. The team swims in the morning and then again for 2 hours in the evening. If she swims 12 hours per week, how long does she swim each morning?**

- Write an algebraic equation to model this problem. Draw a tape diagram to model this problem.
  - *Let $x =$ number of hours of swimming each morning*

    Model days per week (number of hours swimming a.m. and p.m.) $=$ hours of swimming total

$$4 \qquad (x+2) \qquad = 12$$

Recall in the last problem that students used *making zero* first and then *making one* to solve the equation. Explicitly connect *making zero* and *making one* in the previous statement to the tape diagram.

- Solve the equations algebraically and graphically with the help of the tape diagram.
  - *Sample response:*

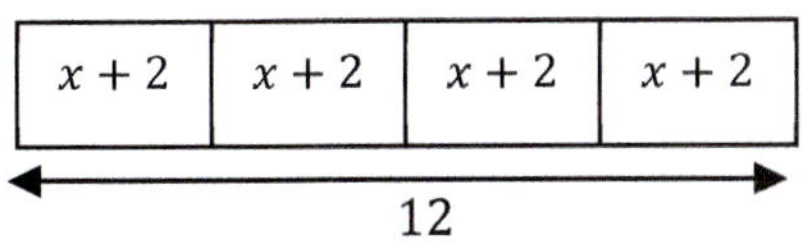	*Algebraically*	
	If: $4\,(x+2) = 12$ *Then:* $\frac{1}{4}\left(4\,(x+2)\right) = \frac{1}{4}(12)$	*Multiplication property of equality using the multiplicative inverse of* 4
$12 - 8 = 4$ $\frac{4}{4} = 1$	*If:* $1(x+2) = 3$ *Then:* $x + 2 = 3$	*Multiplicative identity*
Jenny swims 1 *hour each morning.*	*If:* $x + 2 = 3$ *Then:* $x + 2 - 2 = 3 - 2$	*Subtraction property of equality for the additive inverse of* 2
	If: $x + 0 = 1$ *Then:* $x = 1$	*Additive identity*

- Does your solution make sense in this context? Why?
  - *Yes, if Jenny swims 1 hour in the morning and 2 hours in the evening for a total of 3 hours per day and swims 4 days per week, then $3(4) = 12$ hours for the entire week.*

Exercises (15 minutes)

Exercises

Solve each equation algebraically using if–then statements to justify each step.

1. $5x + 4 = 19$

If: $5x + 4 = 19$

Then: $5x + 4 - 4 = 19 - 4$ — *Subtraction property of equality for the additive inverse of* 4

If: $5x + 0 = 15$

Then: $5x = 15$ — *Additive identity*

If: $5x = 15$

Then: $\frac{1}{5}(5x) = \left(\frac{1}{5}\right)15$ — *Multiplication property of equality for the multiplicative inverse of* 5

If: $1x = 3$

Then: $x = 3$ — *Multiplicative identity*

2. $15x + 14 = 19$

If: $15x + 14 = 19$

Then: $15x + 14 - 14 = 19 - 14$ — *Subtraction property of equality for the additive inverse of* 14

If: $15x + 0 = 5$

Then: $15x = 5$ — *Additive Identity*

If: $15x = 5$

Then: $\frac{1}{15}(15x) = \left(\frac{1}{15}\right)5$ — *Multiplication property of equality for the multiplicative inverse of* 15

If: $1x = \frac{1}{3}$

Then: $x = \frac{1}{3}$ — *Multiplicative identity*

3. **Claire's mom found a very good price on a large computer monitor. She paid $\$325$ for a monitor that was only $\$65$ more than half the original price. What was the original price?**

x: *the original price of the monitor*

If: $\frac{1}{2}x + 65 = 325$

Then: $\frac{1}{2}x + 65 - 65 = 325 - 65$ — *Subtraction property of equality for the additive inverse of* 65

If: $\frac{1}{2}x + 0 = 260$

Then: $\frac{1}{2}x = 260$ — *Additive identity*

If: $\frac{1}{2}x = 260$

Then: $(2)\frac{1}{2}x = (2)260$ — *Multiplication property of equality for the multiplicative inverse of* $\frac{1}{2}$

If: $1x = 520$

Then: $x = 520$ — *Multiplicative identity*

The original price was $\$520$*.*

4. $2(x+4)=18$

If: $2(x+4)=18$

Then: $\frac{1}{2}(2\ (x+4))=\frac{1}{2}(18)$ *Multiplication property of equality using the multiplicative inverse of* 2

If: $1(x+4)=\ 9$

Then: $x+4=9$ *Multiplicative identity*

If: $x+4=9$

Then: $x+4-4=9-4$ *Subtraction property of equality for the additive inverse of* 4

If: $x+0=5$

Then: $x=5$ *Additive identity*

5. **Ben's family left for vacation after his dad came home from work on Friday. The entire trip was 600 mi. Dad was very tired after working a long day and decided to stop and spend the night in a hotel after 4 hours of driving. The next morning, Dad drove the remainder of the trip. If the average speed of the car was 60 miles per hour, what was the remaining time left to drive on the second part of the trip? Remember: Distance = rate multiplied by time.**

m*: the number of miles driven on the second day*

$60\ (m+4)=600$

If: $60\ (m+4)=600$

Then: $\left(\frac{1}{60}\right)60\ (m+4)=\left(\frac{1}{60}\right)600$ *Multiplication property of equality for the multiplicative inverse of* 60

If: $1\ (\ m+4)=10$

Then: $m+4=10$ *Multiplicative identity*

If: $m+4=10$

Then: $m+4-4=10-4$ *Subtraction property of equality for the additive inverse of* 4

If: $m+0=6$

Then: $m=6$ *Additive identity*

There were 6 *hours left to drive.*

Closing (5 minutes)

- What do we mean when we say, "Solve the equation $6x-8=40$?"
  - *Find the value of the variable to make the number sentence true.*
- What property allows us to add 8 to both sides?
  - *Addition property of equality*
- What role does the additive inverse play in solving this equation, and how can you model its use with the tape diagram?
  - *The additive inverse allows us to make zero. This is demonstrated on the tape diagram when we subtract numerical values, and there are no numerical values left.*
- What role does the multiplicative inverse play in solving this equation, and how can you model its use with the tape diagram?
  - *The multiplicative inverse allows us to make one. This is demonstrated on the tape diagram because you can see the number of equal boxes or equal parts.*

- What does this equation look like when modeled using a tape diagram?
  - *Answers will vary because it depends on what type of equation we are modeling.*

Lesson Summary

We work backward to solve an algebraic equation. For example, to find the value of the variable in the equation $6x - 8 = 40$:

1. **Use the addition property of equality to add the opposite of -8 to each side of the equation to arrive at $6x - 8 + 8 = 40 + 8$.**
2. **Use the additive inverse property to show that $-8 + 8 = 0$; thus, $6x + 0 = 48$.**
3. **Use the additive identity property to arrive at $6x = 48$.**
4. **Then use the multiplication property of equality to multiply both sides of the equation by $\frac{1}{6}$ to get:** $\left(\frac{1}{6}\right)6x = \left(\frac{1}{6}\right)48$.
5. **Then use the multiplicative inverse property to show that $\frac{1}{6}(6) = 1$; thus, $1x = 8$.**
6. **Use the multiplicative identity property to arrive at $x = 8$.**

Exit Ticket (5 minutes)

Name ______________________________ Date ______________

Lesson 22: Solving Equations Using Algebra

Exit Ticket

Susan and Bonnie are shopping for school clothes. Susan has $50 and a coupon for a $10 discount at a clothing store where each shirt costs $12.

Susan thinks that she can buy three shirts, but Bonnie says that Susan can buy five shirts. The equations they used to model the problem are listed below. Solve each equation algebraically, justify your steps, and determine who is correct and why.

Susan's Equation	Bonnie's Equation
$12n + 10 = 50$	$12n - 10 = 50$

Exit Ticket Sample Solutions

Susan and Bonnie are shopping for school clothes. Susan has $\$50$ and a coupon for a $\$10$ discount at a clothing store where each shirt costs $\$12$.

Susan thinks that she can buy three shirts, but Bonnie says that Susan can buy five shirts. The equations they used to model the problem are listed below. Solve each equation algebraically, justify your steps, and determine who is correct and why?

Susan's Equation	Bonnie's Equation
$12n + 10 = 50$	$12n - 10 = 50$

Bonnie is correct. The equation that would model this situation is $12n - 10 = 50$. Solving this equation would involve making zero by adding 10. And by doing so, $12n - 10 + 10 = 50 + 10$, we arrive at $12n = 60$. So, if a group of shirts that cost $\$12$ each totals $\$60$, then there must be five shirts since $\frac{60}{12}$ equals 5.

Bonnie's Equation:

$12n - 10 = 50$

$12n - 10 + 10 = 50 + 10$ — *Addition property of equality for the additive inverse of -10*

$12n + 0 = 60$

$12n = 60$ — *Additive identity*

$\left(\frac{1}{12}\right)12n = \left(\frac{1}{12}\right)60$ — *Multiplication property of equality using the multiplicative inverse of 12*

$1n = 5$

$n = 5$ — *Multiplicative identity*

Susan's Equation:

$12n + 10 = 50$

$12n + 10 - 10 = 50 - 10$ — *Subtraction property of equality for the additive inverse of 10*

$12n + 0 = 40$

$12n = 40$ — *Additive identity*

$\left(\frac{1}{12}\right)12n = \left(\frac{1}{12}\right)40$ — *Multiplication property of equality using the multiplicative inverse of 12*

$1n = 3\frac{1}{3}$

$n = 3\frac{1}{3}$ — *Multiplicative identity*

Problem Set Sample Solutions

For each problem below, explain the steps in finding the value of the variable. Then find the value of the variable, showing each step. Write if–then statements to justify each step in solving the equation.

1. $7(m+5) = 21$

   Multiply both sides of the equation by $\frac{1}{7}$, *and then subtract* 5 *from both sides of the equation;* $m = -2$.

If: $7(m+5) = 21$	
Then: $\frac{1}{7}\left(7(m+5)\right) = \frac{1}{7}(21)$	*Multiplication property of equality using the multiplicative inverse of* 7
If: $1\,(m+5) = 3$	
Then: $m+5 = 3$	*Multiplicative identity*
If: $m+5 = 3$	
Then: $m+5-5 = 3-5$	*Subtraction property of equality for the additive inverse of* 5
If: $m+0 = -2$	
Then: $m = -2$	*Additive identity*

2. $-2v+9 = 25$

   Subtract 9 *from both sides of the equation, and then multiply both sides of the equation by* $-\frac{1}{2}$; $v = -8$.

If: $-2v+9 = 25$	
Then: $-2v+9-9 = 25-9$	*Subtraction property of equality for the additive inverse of* 9
If: $-2v+0 = 16$	
Then: $-2v = 16$	*Additive identity*
If: $-2v = 16$	
Then: $-\frac{1}{2}(-2v) = -\frac{1}{2}(16)$	*Multiplication property of equality using the multiplicative inverse of* -2
If: $1\,v = -8$	
Then: $v = -8$	*Multiplicative identity*

3. $\frac{1}{3}y - 18 = 2$

Add 18 to both sides of the equation, and then multiply both sides of the equation by 3; $y = 60$.

If: $\frac{1}{3}y - 18 = 2$

Then: $\frac{1}{3}y - 18 + 18 = 2 + 18$ *Addition property of equality for the additive inverse of* -18

If: $\frac{1}{3}y + 0 = 20$

Then: $\frac{1}{3}y = 20$ *Additive identity*

If: $\frac{1}{3}y = 20$

Then: $3\left(\frac{1}{3}y\right) = 3\,(20)$ *Multiplication property of equality using the multiplicative inverse of* $\frac{1}{3}$

If: $1\,y = 60$

Then: $y = 60$ *Multiplicative identity*

4. $6 - 8p = 38$

Subtract 6 from both sides of the equation, and then multiply both sides of the equation by $-\frac{1}{8}$*;* $p = -4$.

If: $6 - 8p = 38$

Then: $6 - 6 - 8p = 38 - 6$ *Subtraction property of equality for the additive inverse of 6*

If: $0 + (-8p) = 32$

Then: $-8p = 32$ *Additive identity*

If: $-8p = 32$

Then: $\left(-\frac{1}{8}\right)(-8p) = \left(-\frac{1}{8}\right)32$ *Multiplication property of equality using the multiplicative inverse of* -8

If: $1\,p = -4$

Then: $p = -4$ *Multiplicative identity*

5. $15 = 5k - 13$

Add 13 to both sides of the equation, and then multiply both sides of the equation by $\frac{1}{5}$*;* $k = 5.6$.

If: $15 = 5k - 13$

Then: $15 + 13 = 5k - 13 + 13$ *Addition property of equality for the additive inverse of* -13

If: $28 = 5k + 0$

Then: $28 = 5k$ *Additive identity*

If: $28 = 5k$

Then: $\left(\frac{1}{5}\right)28 = \left(\frac{1}{5}\right)5k$ *Multiplication property of equality using the multiplicative inverse of 5*

If: $5.6 = 1k$

Then: $5.6 = k$ *Multiplicative identity*

Lesson 23: Solving Equations Using Algebra

Student Outcomes

- Students use algebra to solve equations (of the form $px + q = r$ and $p(x + q) = r$, where $p, q,$ and r are specific rational numbers) using techniques of *making zero* (adding the additive inverse) and *making one* (multiplying by the multiplicative inverse) to solve for the variable.
- Students identify and compare the sequence of operations used to find the solution to an equation algebraically, with the sequence of operations used to solve the equation with tape diagrams. They recognize the steps as being the same.
- Students solve equations for the value of the variable using inverse operations by *making zero* (adding the additive inverse) and *making one* (multiplying by the multiplicative inverse).

Classwork

MP.1 MP.2 & MP.3

As in Lesson 22, students continue solving equations using properties of equality and inverse operations to relate their steps to the steps taken when solving problems algebraically. In this lesson, students decontextualize word problems to create equations that model given situations. Students justify their solutions by comparing their algebraic steps to the steps taken when using a tape diagram. Have students work in cooperative groups and share out their solutions on chart paper. Use the class discussion as a way to have students view the differences in problem-solving approaches.

Exercises (35 minutes)

Exercises

1. **Youth Group Trip**

   The youth group is going on a trip to an amusement park in another part of the state. The trip costs each group member $\$150$, which includes $\$85$ for the hotel and two one-day combination entrance and meal plan passes.

   a. **Write an equation representing the cost of the trip. Let P be the cost of the park pass.**

      $85 + 2P = 150$

Scaffolding:

Provide a review card showing examples of fraction multiplication and division for students who do not have adequate prerequisite skills.

b. Solve the equation algebraically to find the cost of the park pass. Then write the reason that justifies each step using if-then statements.

If: $85 + 2P = 150$,

Then: $85 - 85 + 2P = 150 - 85$ *Subtraction property of equality for the additive inverse of* 85

If: $0 + 2P = 65$

Then: $2P = 65$ *Additive identity*

If: $2P = 65$

Then: $\left(\frac{1}{2}\right)2P = \left(\frac{1}{2}\right)65$ *Multiplication property of equality using the multiplicative inverse of* 2

If: $1P = 32.5$

Then: $P = 32.5$ *Multiplicative identity*

The park pass costs $\$32.50$.

c. Model the problem using a tape diagram to check your work.

$150 - 85 = 65$

$65 \div 2 = 32.50$

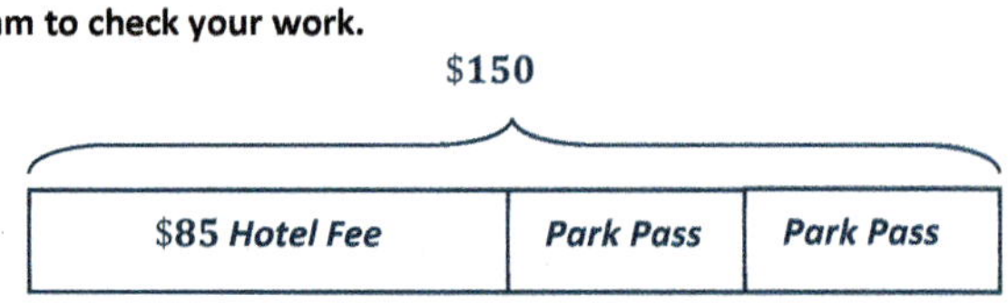

Suppose you want to buy your favorite ice cream bar while at the amusement park, and it costs $\$2.89$. If you purchase the ice cream bar and 3 bottles of water, pay with a $\$10$ bill, and receive no change, then how much did each bottle of water cost?

d. Write an equation to model this situation.

W: *the cost of one bottle of water*

$2.89 + 3W = 10$

e. Solve the equation to determine the cost of one water bottle. Then write the reason that justifies each step using if-then statements.

If: $2.89 + 3W = 10$

Then: $2.89 - 2.89 + 3W = 10 - 2.89$ *Subtraction property of equality for the additive inverse of* 2.89

If: $0 + 3W = 7.11$

Then: $3W = 7.11$ *Additive identity*

If: $3W = 7.11$

Then: $\frac{1}{3}(3W) = \frac{1}{3}(7.11)$ *Multiplication property of equality using the multiplicative inverse of* 3

If: $1W = 2.37$

Then: $W = 2.37$ *Multiplicative identity*

A bottle of water costs $\$2.37$.

f. **Model the problem using a tape diagram to check your work.**

$10 - 2.89 = 7.11$

$\frac{7.11}{3} = 2.37$

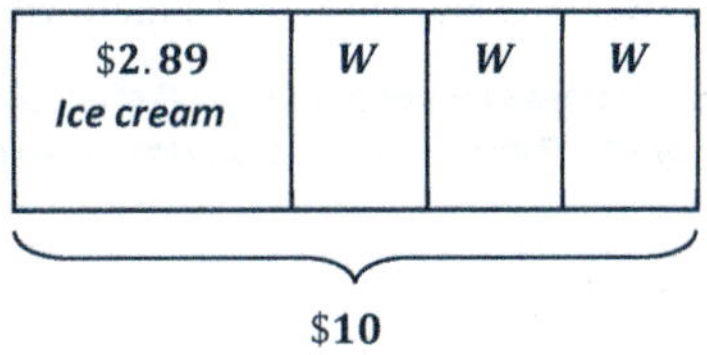

2. **Weekly Allowance**

Charlotte receives a weekly allowance from her parents. She spent half of this week's allowance at the movies but earned an additional $\$4$ for performing extra chores. If she did not spend any additional money and finished the week with $\$12$, what is Charlotte's weekly allowance?

a. **Write an equation that can be used to find the original amount of Charlotte's weekly allowance. Let A be the value of Charlotte's original weekly allowance.**

$\frac{1}{2}A + 4 = 12$

b. **Solve the equation to find the original amount of allowance. Then write the reason that justifies each step using if-then statements.**

If: $\frac{1}{2}A + 4 = 12$

Then: $\frac{1}{2}A + 4 - 4 = 12 - 4$ — *Subtraction property of equality for the additive inverse of* 4

If: $\frac{1}{2}A + 0 = 8$

Then: $\frac{1}{2}A = 8$ — *Additive identity*

If: $\frac{1}{2}A = 8$

Then: $(2)\frac{1}{2}A = (2)8$ — *Multiplication property of equality using the multiplicative inverse of* $\frac{1}{2}$

If: $1A = 16$

Then: $A = 16$ — *Multiplicative identity*

The original allowance was $\$16$*.*

c. **Explain your answer in the context of this problem.**

Charlotte's weekly allowance is $\$16$*.*

d. **Charlotte's goal is to save $\$100$ for her beach trip at the end of the summer. Use the amount of weekly allowance you found in part (c) to write an equation to determine the number of weeks that Charlotte must work to meet her goal. Let w represent the number of weeks.**

$$16w = 100$$
$$\left(\frac{1}{16}\right)16w = \left(\frac{1}{16}\right)100$$
$$1w = 6.25$$
$$w = 6.25$$

e. In looking at your answer to part (d) and based on the story above, do you think it will take Charlotte that many weeks to meet her goal? Why or why not?

Charlotte needs more than 6 weeks' allowance, so she will need to save 7 weeks' allowance (and not spend any of it). There are 10–12 weeks in the summer; so, yes, she can do it.

3. Travel Baseball Team

Allen is very excited about joining a travel baseball team for the fall season. He wants to determine how much money he should save to pay for the expenses related to this new team. Players are required to pay for uniforms, travel expenses, and meals.

a. If Allen buys 4 uniform shirts at one time, he gets a $\$10.00$ discount so that the total cost of 4 shirts would be $\$44$. Write an algebraic equation that represents the regular price of one shirt. Solve the equation. Write the reason that justifies each step using if-then statements.

s: the cost of one shirt

If: $4s - 10 = 44$

Then: $4s - 10 + 10 = 44 + 10$ — *Addition property of equality using the additive inverse of* -10

If: $4s + 0 = 54$

Then: $4s = 54$ — *Additive identity*

If: $4s = 54$

Then: $\left(\frac{1}{4}\right)4s = \left(\frac{1}{4}\right)54$ — *Multiplication property of equality using multiplicative inverse of* 4

If: $1s = 13.50$

Then: $s = 13.50$ — *Multiplicative identity*

b. What is the cost of one shirt without the discount?

The cost of one shirt is $\$13.50$.

c. What is the cost of one shirt with the discount?

$$4s = 44$$
$$\left(\frac{1}{4}\right)4s = \left(\frac{1}{4}\right)44$$
$$1s = 11$$
$$s = 11$$

The cost of one shirt with the discount is $\$11.00$.

d. How much more do you pay per shirt if you buy them one at a time (rather than in bulk)?

$13.50 - 11.00 = 2.50$

One shirt costs $\$11$ *if you buy them in bulk. So, Allen would pay* $\$2.50$ *more per shirt if he bought them one at a time.*

Allen's team was also required to buy two pairs of uniform pants and two baseball caps, which total $\$68$. A pair of pants costs $\$12$ more than a baseball cap.

e. Write an equation that models this situation. Let c represent the cost of a baseball cap.

$2(\textit{cap} + 1\ \textit{pair of pants}) = 68$

$2(c + c + 12) = 68$ *or* $2(2c + 12) = 68$ *or* $4c + 24 = 68$

f. **Solve the equation algebraically to find the cost of a baseball cap. Write the reason that justifies each step using if-then statements.**

If: $2\ (2\ c + 12) = 68$

Then: $\left(\frac{1}{2}\right)(2\ )(2\ c + 12) = \left(\frac{1}{2}\right) 68$ — *Multiplication property of equality using the multiplicative inverse of* 2

If: $1(2c + 12) = 34$

Then: $2c + 12 = 34$ — *Multiplicative identity*

If: $2c + 12 = 34$

Then: $2c + 12 - 12 = 34 - 12$ — *Subtraction property of equality for the additive inverse of* 12

If: $2c + 0 = 22$

Then: $2c = 22$ — *Additive identity*

If: $2c = 22$

Then: $\left(\frac{1}{2}\right) 2c = \left(\frac{1}{2}\right) 22$ — *Multiplication property of equality using the multiplicative inverse of* 2

If: $1c = 11$

Then: $c = 11$ — *Multiplicative identity*

g. **Model the problem using a tape diagram in order to check your work from part (f).**

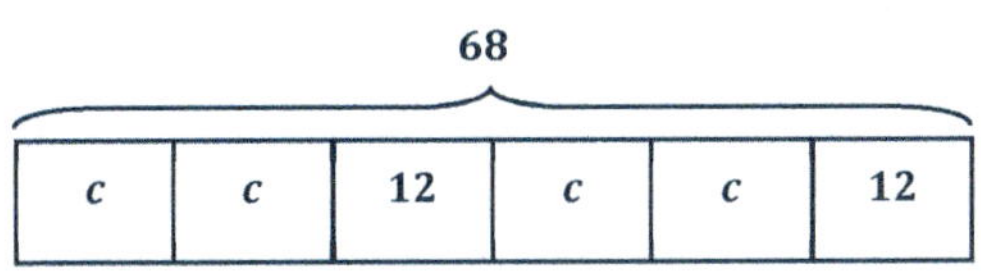

$68 - 24 = 44$

$\frac{44}{4} = 11$

h. **What is the cost of one cap?**

The cost of one cap is $\$11$.

i. **What is the cost of one pair of pants?**

$11 + 12 = 23$ — *The cost of one pair of pants is* $\$23$.

Closing (5 minutes)

- How do we translate a word problem into an equation? For instance, in Exercise 1 about the youth group trip, what key words and statements helped you determine the operations and values used in the equation?
  - *Answers will vary, but students should talk about key words to determine the total value, the constant value, and the coefficient.*
- How do we make sense of a word problem and model it with an equation?
  - *Answers will vary and may be similar to answers to the previous question.*

> **Lesson Summary**
>
> **Equations are useful to model and solve real-world problems. The steps taken to solve an algebraic equation are the same steps used in an arithmetic solution.**

Exit Ticket (5 minutes)

Name ____ Date ____________________

Lesson 23: Solving Equations Using Algebra

Exit Ticket

Andrew's math teacher entered the seventh-grade students in a math competition. There was an enrollment fee of $\$30$ and also an $\$11$ charge for each packet of 10 tests. The total cost was $\$151$. How many tests were purchased?

Set up an equation to model this situation, solve it using if-then statements, and justify the reasons for each step in your solution.

Exit Ticket Sample Solutions

Andrew's math teacher entered the seventh-grade students in a math competition. There was an enrollment fee of $\$30$ and also an $\$11$ charge for each packet of 10 tests. The total cost was $\$151$. How many tests were purchased?

Set up an equation to model this situation, solve it using if-then statements, and justify the reasons for each step in your solution.

Let p represent the number of test packets.

Enrollment fee + cost of test = 151

If: $30 + 11p = 151$

Then: $30 - 30 + 11p = 151 - 30$ — *Subtraction property of equality for the additive inverse of* 30

If: $0 + 11p = 121$

Then: $11p = 121$ — *Additive identity*

If: $11p = 121$

Then: $\frac{1}{11}(11p) = \frac{1}{11}(121)$ — *Multiplication property of equality using the multiplicative inverse of* 11

If: $1p = 11$

Then: $p = 11$ — *Multiplicative identity*

Andrew's math teacher bought 11 *packets of tests. There were* 10 *tests in each packet, and* $10 \times 11 = 110$.

So, there were 110 *tests purchased.*

Problem Set Sample Solutions

For Exercises 1–4, solve each equation algebraically using if-then statements to justify your steps.

1. $\frac{2}{3}x - 4 = 20$

If: $\frac{2}{3}x - 4 = 20$

Then: $\frac{2}{3}x - 4 + 4 = 20 + 4$ — *Addition property of equality using the additive inverse of* -4

If: $\frac{2}{3}x + 0 = 24$

Then: $\frac{2}{3}x = 24$ — *Additive identity*

If: $\frac{2}{3}x = 24$

Then: $\left(\frac{3}{2}\right)\frac{2}{3}x = \left(\frac{3}{2}\right)24$ — *Multiplication property of equality using the multiplicative inverse of* $\frac{2}{3}$

If: $1x = 36$

Then: $x = 36$ — *Multiplicative identity*

2. $4 = \frac{-1+x}{2}$

If: $4 = \frac{-1+x}{2}$

Then: $2\,(4) = 2\left(\frac{-1+x}{2}\right)$ — *Multiplication property of equality using the multiplicative inverse of* $\frac{1}{2}$

If: $8 = 1\,(-1+x)$

Then: $8 = -1+x$ — *Multiplicative identity*

If: $8 = -1+x$

Then: $8-(-1) = -1-(-1)+x$ — *Subtraction property of equality for the additive inverse of* -1

If: $9 = 0+x$

Then: $9 = x$ — *Additive identity*

3. $12(x+9) = -108$

If: $12(x+9) = -108$

Then: $\left(\frac{1}{12}\right)12(x+9) = \left(\frac{1}{12}\right)(-108)$ — *Multiplication property of equality using the multiplicative inverse of* 12

If: $1\,(x+9) = -9$

Then: $x+9 = -9$ — *Multiplicative identity*

If: $x+9 = -9$

Then: $x+9-9 = -9-9$ — *Subtraction property of equality for the additive inverse of* 9

If: $x+0 = -18$

Then: $x = -18$ — *Additive identity*

4. $5x + 14 = -7$

If: $5x + 14 = -7$

Then: $5x + 14 - 14 = -7 - 14$ — *Subtraction property of equality for the additive inverse of* 14

If: $5x + 0 = -21$

Then: $5x = -21$ — *Additive identity*

If: $5x = -21$

Then: $\left(\frac{1}{5}\right)5x = \left(\frac{1}{5}\right)(-21)$ — *Multiplication property of equality using the multiplicative inverse of* 5

If: $1x = -4.2$

Then: $x = -4.2$ — *Multiplicative identity*

For Exercises 5–7, write an equation to represent each word problem. Solve the equation showing the steps, and then state the value of the variable in the context of the situation.

5. A plumber has a very long piece of pipe that is used to run city water parallel to a major roadway. The pipe is cut into two sections. One section of pipe is 12 ft. shorter than the other. If $\frac{3}{4}$ of the length of the shorter pipe is 120 ft., how long is the longer piece of the pipe?

   Let x represent the longer piece of pipe.

   If: $\frac{3}{4}(x-12)=120$

   Then: $\frac{4}{3}\left(\frac{3}{4}\right)(x-12)=\left(\frac{4}{3}\right)120$ *Multiplication property of equality using the multiplicative inverse of* $\frac{3}{4}$

   If: $1(x-12)=160$

   Then: $x-12=160$ *Multiplicative identity*

   If: $x-12=160$

   Then: $x-12+12=160+12$ *Addition property of equality for the additive inverse of* -12

   If: $x+0=172$

   Then: $x=172$ *Additive identity*

   The longer piece of pipe is 172 *ft.*

6. Bob's monthly phone bill is made up of a $\$10$ fee plus $\$0.05$ per minute. Bob's phone bill for July was $\$22$. Write an equation to model the situation using m to represent the number of minutes. Solve the equation to determine the number of phone minutes Bob used in July.

   Let m represent the number of phone minutes Bob used.

   If: $10+0.05m=22$

   Then: $10-10+0.05m=22-10$ *Subtraction property of equality for the additive inverse of* 10

   If: $0+0.05m=12$

   Then: $0.05m=12$ *Additive identity*

   If: $0.05m=12$

   Then: $\left(\frac{1}{0.05}\right)0.05m=\left(\frac{1}{0.05}\right)12$ *Multiplication property of equality using the multiplicative inverse of* 0.05

   If: $1m=240$

   Then: $m=240$ *Multiplicative identity*

   Bob used 240 *phone minutes in July.*

7. **Kym switched cell phone plans. She signed up for a new plan that will save her $\$3.50$ per month compared to her old cell phone plan. The cost of the new phone plan for an entire year is $\$294$. How much did Kym pay per month under her old phone plan?**

Let n represent the amount Kym paid per month for her old cell phone plan.

If: $294 = 12(n - 3.50)$

Then: $\left(\frac{1}{12}\right)(294) = \left(\frac{1}{12}\right)12(n - 3.50)$ — *Multiplication property of equality using the multiplicative inverse of* 12

If: $24.5 = 1\,(n - 3.50)$

Then: $24.5 = n - 3.50$ — *Multiplicative identity*

If: $24.5 = n - 3.50$

Then: $24.5 + 3.50 = n - 3.50 + 3.50$ — *Addition property of equality for the additive inverse of* -3.50

If: $28 = n + 0$

Then: $28 = n$ — *Additive identity*

Kym paid $\$28$ per month for her old cell phone plan.

Name ______________________________ Date ______________

1. The water level in Ricky Lake changes at an average of $-\frac{7}{16}$ inch every 3 years.

   a. Based on the rate above, how much will the water level change after one year? Show your calculations and model your answer on the vertical number line, using 0 as the original water level.

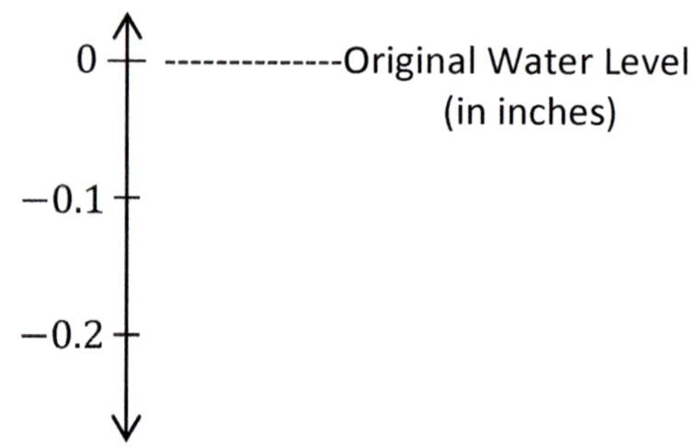

   b. How much would the water level change over a 7-year period?

   c. When written in decimal form, is your answer to part (b) a repeating decimal or a terminating decimal? Justify your answer using long division.

2. Kay's mother taught her how to make handmade ornaments to sell at a craft fair. Kay rented a table at the fair for $\$30$ and set up her work station. Each ornament that she makes costs approximately $\$2.50$ for materials. She sells each ornament for $\$6.00$.

   a. If x represents the number of ornaments sold at the craft fair, which of the following expressions would represent Kay's profit? (Circle *all* choices that apply.)

      A. $-30 + 6x - 2.50x$

      B. $6x - 30 - 2.50x$

      C. $6x - 30$

      D. $4.50x - 30$

      E. $3.50x - 30$

   b. Kay does not want to lose money on her business. Her mother told her she needs to sell enough ornaments to at least cover her expenses (costs for materials and table rental). Kay figures that if she sells 8 ornaments, she covers her expenses and does not lose any money. Do you agree? Explain and show work to support your answer.

   c. Kay feels that if she earns a profit of $\$40.00$ at this craft fair, her business will be successful enough for her to branch out to other craft fairs. How many ornaments does she have to sell to earn a $\$40.00$ profit? Write and solve an equation; then explain how the steps and operations used in your algebraic solution compare to an arithmetic solution.

3. Travis received a letter from his bank saying that his checking account balance fell below zero. His account transaction log is shown below.

CHECK NO.	DATE	DESCRIPTION OF TRANSACTION	PAYMENT	DEPOSIT	BALANCE	
---	10/17	Beginning Balance	---	---	$367.50	
1125	10/18	CBC Audio (Headphones)	$62.00		−62.00	
					$305.50	Line 1
1126	10/22	NY Sport (Basketball Shoes)	$87.00		−87.00	
					$218.50	Line 2
Debit	10/25	Gary's Country Market	$38.50		−38.50	
					$180.00	Line 3
1127	10/25	Iggy's Skate Shop (Skateboard)	$188.00		−188.00	
					$8.00	Line 4
	10/25	Cash Deposit (Birthday Money)		$20.00	+20.00	
					$28.00	Line 5
Debit	10/30	McDonuts	$5.95		−5.95	
					$22.05	Line 6

a. On which line did Travis make a mathematical error? Explain Travis's mistake.

b. The bank charged Travis a $20 fee because his balance dropped below $0. He knows that he currently has an outstanding charge for $7.85 that he has not recorded yet. How much money will Travis have to deposit into his account so that the outstanding charge does not create another bank fee? Explain.

4. The length of a rectangular envelope is $2\frac{1}{2}$ times its width. A plastic band surrounds the front and back of the envelope to secure it as shown in the picture. The plastic band is $39\frac{3}{8}$ inches long. Find the length and width of the envelope.

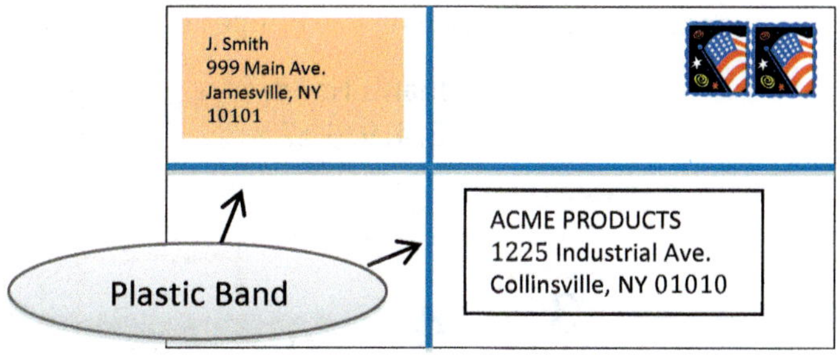

5. Juan and Mary are playing the Integer Card Game. The cards in their hands are shown below:

Juan's Hand
$3, 4, 9, -12$

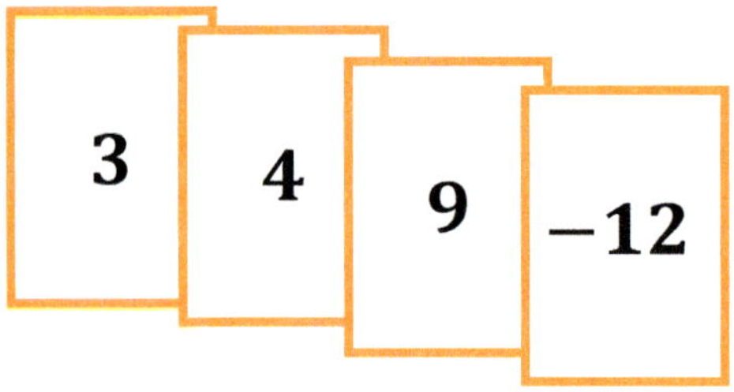

Mary's Hand
$-2, 3, 1, 2$

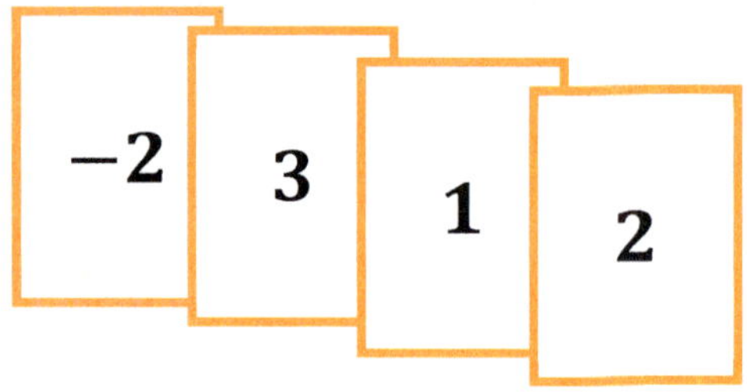

a. What are the scores in each of their hands?

Juan's score:

Mary's score:

b. Lydia says that if Juan and Mary both take away their 3's, Juan's score will be higher than Mary's. Marcus argues and says that Juan and Mary's scores will be equal. Are either of them right? Explain.

c. Juan picks up another set of cards that is exactly like each card in his hand. Which of the following would make Mary's score equal to Juan's? Place a check mark ✓ by all that apply.

_____Double every card in her hand

_____Take away her 3 and 1

_____Pick up a 4

_____Take away her 2 and -2

_____Pick up a 7 and -3

_____Pick up one of each of Juan's cards

Explain why your selections will make Juan's and Mary's scores equal.

A Progression Toward Mastery					
Assessment Task Item		**STEP 1** Missing or incorrect answer and little evidence of reasoning or application of mathematics to solve the problem.	**STEP 2** Missing or incorrect answer but evidence of some reasoning or application of mathematics to solve the problem.	**STEP 3** A correct answer with some evidence of reasoning or application of mathematics to solve the problem, OR an incorrect answer with substantial evidence of solid reasoning or application of mathematics to solve the problem.	**STEP 4** A correct answer supported by substantial evidence of solid reasoning or application of mathematics to solve the problem.
1	**a** 7.NS.A.2b	Student incorrectly calculates the water level change with either no model shown or the model shown does not relate to the answer given.	Student sets up the problem correctly but makes an error in computation resulting in an incorrect value and incorrectly models the answer.	Student uses a sound process to determine and model the answer on the number line, but a computational error results in an incorrect value. OR Student correctly calculates a change of $-\frac{7}{48}$ inches but has an error in the number line representation.	Student correctly states that the water level changes $-\frac{7}{48}$ inches after one year and correctly models the change on the number line.
	b 7.NS.A.2a	Student answer is incorrect. Student work shows little or no understanding of how to find the water level change over a 7-year period.	Student uses an appropriate method to find the water level change, but a computational error results in an incorrect value. Student does not correctly interpret that value to describe the change.	Student uses an appropriate method to find and express the 7-year water level change, but a computational error results in an incorrect value. OR Student states a change of $1\frac{1}{48}$ inches but does not indicate the sign or direction of that change.	Student correctly states the $-1\frac{1}{48}$ inch change in the water level over a 7-year period and uses an appropriate method to obtain the answer.

	c **7.NS.A.2d**	Student is unable to demonstrate correct use of the long division algorithm.	Student shows partial understanding of the long division algorithm but does not complete the process.	Student uses long division to determine and justify the decimal form of the answer, but a computational error results in an incorrect value. OR Student shows the correct long division work to arrive at a decimal remainder but does not use a repeat bar to indicate a repeat pattern.	Student correctly uses the long division algorithm to determine that $1\frac{1}{48}$ is the repeating decimal $1.0208\overline{3}$ (or that $-1\frac{1}{48}$ equals $-1.0208\overline{3}$). OR Student uses the long division algorithm to correctly determine and state the decimal form of a different answer that is recorded in part (b).
2	**a** **7.EE.A.2**	Student does not circle any of A, B, or E. OR Student circles only one of A, B, and E and circles C and/or D. OR Student circles all choices. OR Student does not circle any choices.	Student circles only two of A, B, and E and also circles C or D. OR Student circles only one of A, B, and E.	Student circles only two out of A, B, and E.	Student circles only choices A, B, and E.
	b **7.NS.A.3**	Student shows some accuracy in mathematical computation, but the work is not relevant. Student fails to provide an explanation or provides an incorrect explanation.	Student arrives at a value of -2 for the amount of money Kay made from selling 8 ornaments but incorrectly agrees with the claim. OR Student does not make a statement to agree or disagree.	Student arrives at a value of -2 for the amount of money Kay made from selling 8 ornaments and disagrees with the statement but does not provide a complete explanation. OR Due to a minor computational error, student arrives at an incorrect answer but includes a sound explanation based on that numerical answer.	Student correctly disagrees with the statement and supports the answer with the appropriate work. For instance, student shows that $3.50(8) - 30 = -2$, which means Kay would have lost $2.

	c **7.NS.A.3** **7.EE.B.4a**	Student answer is incorrect. Little or no evidence of reasoning is provided.	Student answer is incorrect but shows some evidence of reasoning through the use of an equation and/or arithmetic steps to model and solve the problem (though the model used may be incorrect).	Student uses a correct equation and method (e.g., $3.50x - 30 = 40$ and finds 20 to be the number of ornaments Kay must sell) but does not provide an explanation for how the steps or solution compares to an arithmetic solution. OR Student uses a correct equation and method and relates it to an arithmetic model but makes a computational error resulting in an incorrect value.	Student correctly states that Kay must sell 20 ornaments to earn a $\$40$ profit and includes a correct equation and relates the steps in the solution to an arithmetic model with no errors in the steps taken to arrive at the answer.
3	**a** **7.NS.A.1**	Student does not provide a correct explanation. Student identifies a different line and shows little or no evidence of understanding integer subtraction.	Student correctly identifies line 4 but does not explain the mistake or state a correct value for line 4. OR Student makes an error in computation and states an incorrect value for line 4. OR Student identifies another line as being Travis's mistake, due to a computational error, but shows an understanding of integer subtraction.	Student correctly identifies line 4 and states that the value should instead be $-\$8$ but does not clearly explain the mistake. OR Student clearly explains the mistake but does not provide the correct value.	Student correctly identifies line 4, states that Travis mistakenly obtained a positive difference from $180 - 188$ and states that the value on line 4 should instead be $-\$8$.

	b **7.NS.A.1**	Student is unable to answer the question accurately. Student makes several errors in calculating the correct account balance and necessary deposit, which shows a limited level of understanding.	Student uses an incorrect beginning balance (such as $\$22.05$ from line 6) to calculate the new account balance but performs all other calculations correctly and explains that the account balance needs to be at least $\$0$. OR Student corrects Travis's initial error and arrives at an account balance of $\$6.05$ but does not complete the other necessary steps to determine the deposit needed.	Student answers incorrectly due to a computational error but uses a sound process and valid explanation of how much Travis should deposit into the account (based on the incorrect value). OR Student shows a correct process and arrives at a new balance amount of $-\$21.80$ but does not provide a complete explanation of how much money Travis needed to deposit.	Student calculates the correct account balance of $-\$21.80$ showing appropriate work, states the need for a deposit of $\$21.80$ or more to avoid overdraft and explains that the deposit is necessary to reach a balance of at least $\$0$.
4	**7.NS.A.3** **7.EE.B.4a**	Student answers incorrectly and shows little or no understanding of how to find the missing dimensions of the envelope.	Student uses a valid process to arrive at either a correct length of $14\frac{1}{16}$ inches or width of $5\frac{5}{8}$ inches but does not provide both dimensions. OR Student relates the length and width backward, resulting in a length of $5\frac{5}{8}$ inches and a width of $14\frac{1}{16}$ inches.	Student provides appropriate work and correct numerical values for the answer but without the units of measure. OR Student provides incorrect answer values based on a computational error but uses a valid method (i.e., $2w + 5w = 39.375$) and shows correct steps.	Student correctly answers a length of $14\frac{1}{16}$ inches and width of $5\frac{5}{8}$ inches and provides error-free work to support the answer.
5	**a** **7.NS.A.1**	Student is unable to correctly answer the question. Student work is missing or does not demonstrate an adequate understanding of integer addition.	Student correctly indicates that Juan and Mary each have scores of 4 but does not show supporting work.	Student calculates and shows that one of the scores is 4, but for the other hand, a computational error is made resulting in a different value.	Student correctly calculates and shows that Juan and Mary each have scores of 4.
	b **7.NS.A.1**	Student states that Lydia is correct OR Student states that neither person is correct.	Student states that Marcus is correct but provides no explanation as to why.	Student states that Marcus is correct, but the explanation is incomplete.	Student states that Marcus is correct and provides a valid argument as justification.

	c 7.NS.A.3	Student checks both of the incorrect choices, and the written explanation shows little or no understanding.	Student places check marks by only two of the correct choices (and possibly one of the incorrect choices). Student explanation indicates a limited level of understanding.	Student provides all but one of the following: ▪ Student places check marks by only the four correct answers; ▪ Student explains that Juan's score is 8 because it was doubled; ▪ Student explains why the selections will make the scores equal. OR Student checks only 3 of the 4 correct choices but appropriately addresses all other parts of the question.	Student places check marks by only the four correct answers and explains that Juan's score is 8 because it was doubled and accurately explains why the selections will make the scores equal.

G7-M2-B2-v1.3.0_05/04/15

Name ________________________________ Date ________________

1. The water level in Ricky Lake changes at an average of $-\frac{7}{16}$ inch every 3 years.

   a. Based on the rate above, how much will the water level change after one year? Show your calculations and model your answer on the vertical number line, using 0 as the original water level.

$\frac{-7}{13} \div 3 \Rightarrow \frac{-7}{13} \cdot \frac{1}{3} \Rightarrow \frac{-7}{48}$

```
       0.145
48 | 7.0000
   - 48
      220
    - 192
       280
     - 240
         40
```

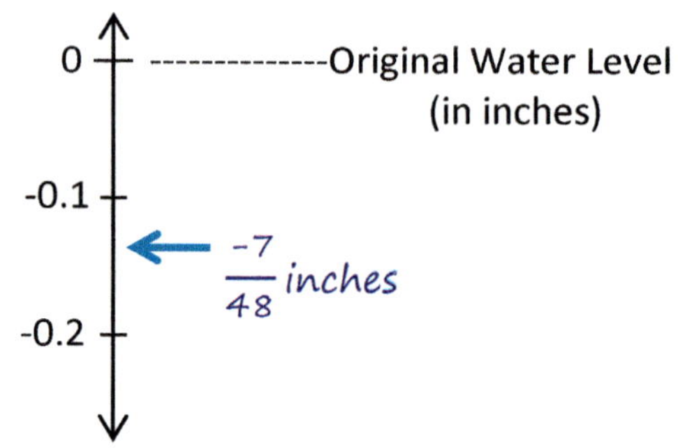

   b. How much would the water level change over a 7-year period?

Distance = rate × time

$= -\frac{7}{48} \times 7$

$= -\frac{49}{48}$

$= -1\frac{1}{48}$

The water level drops $1\frac{1}{48}$ inches over a 7-year period.

   c. When written in decimal form, is your answer to part (b) a repeating decimal or a terminating decimal? Justify your answer using long division.

```
     0.020833
48 | 1.000000
   -  96
       40
    -  00
       400
    -  384
        160
     -  144
        160
```

$-1\frac{1}{48}$ written in decimal form is a repeating decimal because when converted using long division, the remainder repeats after the hundred-thousandths place.

2. Kay's mother taught her how to make handmade ornaments to sell at a craft fair. Kay rented a table at the fair for $30 and set up her work station. Each ornament that she makes costs approximately $2.50 for materials. She sells each ornament for $6.00.

   a. If x represents the number of ornaments sold at the craft fair, which of the following expressions would represent Kay's profit? (Circle *all* choices that apply.)

      (A.) $-30 + 6x - 2.50x$
      (B.) $6x - 30 - 2.50x$
      C. $6x - 30$
      D. $4.50x - 30$
      (E.) $3.50x - 30$

   b. Kay does not want to lose money on her business. Her mother told her she needs to sell enough ornaments to at least cover her expenses (costs for materials and table rental). Kay figures that if she sells 8 ornaments, she covers her expenses and does not lose any money. Do you agree? Explain and show work to support your answer.

$$3.50x - 30$$
$$3.50(8) - 30$$
$$(24 + 4) - 30$$
$$28 - 30$$
$$-2$$

I disagree with Kay because selling 8 ornaments covers most of her costs but still leaves her $2 in debt.

   c. Kay feels that if she earns a profit of $40.00 at this craft fair, her business will be successful enough for her to branch out to other craft fairs. How many ornaments does she have to sell to earn a $40.00 profit? Write and solve an equation; then explain how the steps and operations used in your algebraic solution compare to an arithmetic solution.

$$3.50x - 30 = 40$$
$$3.50x - 30 + 30 = 40 + 30$$
$$3.50x + 0 = 70$$
$$3.50x = 70$$
$$3.50x\left(\frac{1}{3.50}\right) = 70\left(\frac{1}{3.50}\right)$$
$$1x = 20$$

Kay must sell 20 ornaments.

To find the answer arithmetically, I would have to combine the $40 profit and $30 rental fee, then divide that sum ($70) by the $3.50 that she earns per ornament after costs.

3. Travis received a letter from his bank saying that his checking account balance fell below zero. His account transaction log is shown below.

CHECK NO.	DATE	DESCRIPTION OF TRANSACTION	PAYMENT	DEPOSIT	BALANCE	
---	10/17	Beginning Balance	---	---	$367.50	
1125	10/18	CBC Audio (Headphones)	$62.00		−62.00	
					$305.50	Line 1
1126	10/22	NY Sport (Basketball Shoes)	$87.00		−87.00	
					$218.50	Line 2
Debit	10/25	Gary's Country Market	$38.50		−38.50	
					$180.00	Line 3
1127	10/25	Iggy's Skate Shop (Skateboard)	$188.00		−188.00	
					$8.00	Line 4
	10/25	Cash Deposit (Birthday Money)		$20.00	+20.00	
					$28.00	Line 5
Debit	10/30	McDonuts	$5.95		−5.95	
					$22.05	Line 6

a. On which line did Travis make a mathematical error? Explain Travis's mistake.

On line 4, Travis subtracted $188 from $180 and got a positive answer. The difference should be -$8.00.

b. The bank charged Travis a $20 fee because his balance dropped below 0. He knows that he currently has an outstanding charge for $7.85 that he has not recorded yet. How much money will Travis have to deposit into his account so that the outstanding charge does not create another bank fee? Explain.

Starting at Line 3:

$180.00 - 188.00 + 20.00 - 5.95$

$-8.00 + 20.00 - 5.95$

$12.00 - 5.95$

6.05

Travis's actual balance should be $6.05.

$6.05 + (-20.00)$ *overdraft fee*

$|-20.00| - |6.05|$

-13.95

$-13.95 + (-7.85)$ *outstanding charge*

$|-13.95| + |-7.85|$

-21.80

To get his account back to 0, Travis needs to deposit $21.80 or more to avoid another overdraft fee.

4. The length of a rectangular envelope is $2\frac{1}{2}$ times its width. A plastic band surrounds the front and back of the envelope to secure it as shown in the picture. The plastic band is $39\frac{3}{8}$ inches long. Find the length and width of the envelope.

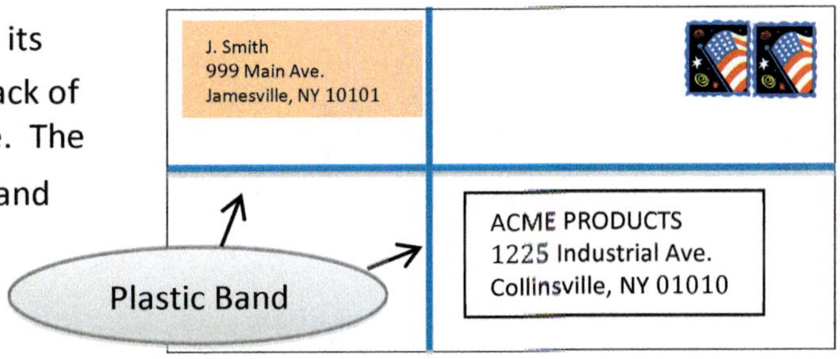

The length of the plastic band is equivalent to the perimeter of the envelope.

Width: w

Length $= 2\frac{1}{2} \times$ *width*

$$w + w + \left(2\frac{1}{2}w\right) + \left(2\frac{1}{2}w\right) = 39\frac{3}{8}$$

$$2w + 5w = 39\frac{3}{8}$$

$$7w = 39\frac{3}{8}$$

$$7w = \frac{315}{8}$$

$$\left(\frac{1}{7}\right)(7w) = \left(\frac{315}{8}\right)\left(\frac{1}{7}\right)$$

$$w = \frac{45}{8} = 5\frac{5}{8}$$

$$\text{Length} = 2\frac{1}{2}w$$

$$\text{Length} = 2\frac{1}{2}\left(5\frac{5}{8}\right)$$

$$\text{Length} = \frac{5}{2} \times \frac{45}{8}$$

$$\text{Length} = \frac{225}{16} = 14\frac{1}{16}$$

The length of the envelope is $14\frac{1}{16}$ *inches and the width is* $5\frac{5}{8}$ *inches.*

5. Juan and Mary are playing the Integer Card Game. The cards in their hands are shown below:

Juan's Hand
$3, 4, 9, -12$

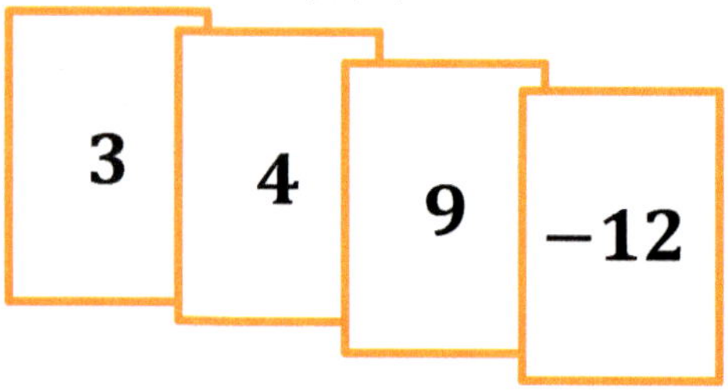

Mary's Hand
$-2, 3, 1, 2$

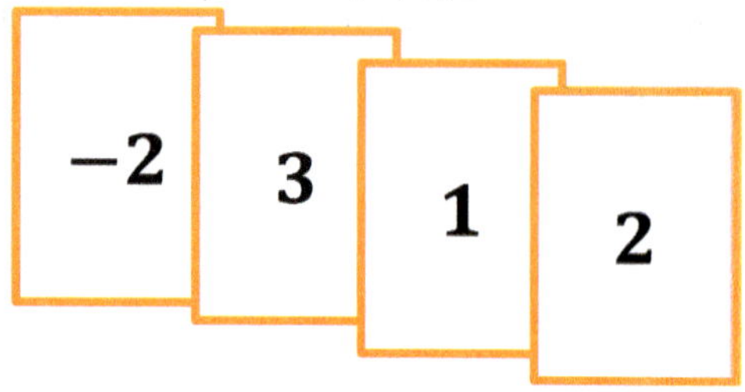

a. What are the scores in each of their hands?

Juan's Score: 4
$3 + 4 + 9 + (-12)$
$7 + 9 + (-12)$
$16 + (-12)$
4

Mary's Score: 4
$-2 + 3 + 1 + 2$
$1 + 1 + 2$
4

b. Lydia says that if Juan and Mary both take away their 3's, Juan's score will be higher than Mary's. Marcus argues and says that Juan and Mary's scores will be equal. Are either of them right? Explain.

If both Juan and Mary lay down their 3's, then both of their totals will be decreased by 3. Since both of their totals are 4, laying down a 3 would make both scores 1. Juan's score and Mary's score would be equal, so Marcus is correct.

c. Juan picks up another set of cards that is exactly like each card in his hand. Which of the following would make Mary's score equal to Juan's? Place a check mark ✓ by all that apply.

__✓__ Double every card in her hand
+ total of 4

_____ Take away her 3 and 1

__✓__ Pick up a 4
+ 4

_____ Take away her 2 and −2

__✓__ Pick up a 7 and −3
+ 4

__✓__ Pick up one of each of Juan's cards
+ 4

Explain why your selections will make Juan and Mary's scores equal.

Juan's total doubles because every card in his hand doubled, so his total is 8. Each choice I selected would add 4 to Mary's total to make it 8.